Engineering Instrumentation and Control

Engineering Instrumentation and Control

W. Bolton

Butterworth-Heinemann Ltd
Linacre House, Jordan Hill, Oxford OX2 8DP

A member of the Reed Elsevier plc group

OXFORD LONDON BOSTON
NEW DELHI SINGAPORE SYDNEY
TOKYO TORONTO WELLINGTON

First published 1996

British Library Cataloguing in Publication Data
Bolton, W.
1 Engineering instrumentation and control
I. Title

ISBN 0 7506 2725 5

Library of Congress Cataloging in Publication Data
A catalogue record for this book is available from the Library of Congress

Printed by Martin's The Printers Ltd, Berwick upon Tweed

Contents

Preface

The aim of this book is to enable the reader to:

- Identify the functional elements of measurement systems and describe the operating principles and characteristics of common elements.
- Describe and explain the operation of examples of measurement systems used for common engineering measurements.
- Maintain and test engineering measurement systems.
- Identify the different types of control systems and their functional elements.
- Describe controllers used in engineering control systems, the different control modes, programmable logic controllers and ladder diagrams.
- Describe final control elements used in engineering control systems and their operating principles and characteristics.
- Describe and explain the operation of examples of process, servo and programmable logic controlled systems.

This book has been written to provide a comprehensive coverage of the Engineering Instrumentation and Control unit for the Advanced GNVQ in Engineering, including revision of principles covered in earlier GNVQ units, and also to provide a general introduction to that topic for those taking other engineering courses. The aims of the chapters and their relationship to the elements of the GNVQ unit, though the chapters often include revision of earlier concepts and give a wider coverage, are:

Chapter	*Aim*	*GNVQ element*
1	Identify the functional elements and requirements of measurement systems	14.1
2	Describe the operating principles and characteristics of functional elements in engineering measurement systems.	14.1
3	Describe and explain the operation of examples of measurement systems used for common engineering measurements.	14.1
4	Maintain and test engineering measurement systems.	14.1
5	Identify the different types of control systems and their functional elements.	14.2

6	Describe controllers used in engineering control systems, the different control modes, programmable logic controllers and ladder diagrams.	14.2
7	Describe final control elements used in engineering control systems and their operating principles and characteristics.	14.3
8	Describe and explain the operation of examples of process, servo and programmable logic controlled systems.	14.2 14.3

At the end of each chapter there are multiple-choice questions and problems. Answers are given for all the multiple-choice questions and guidance given as to the answers for all the problems.

W. Bolton

1 Measurement systems

1.1 Measurement systems

The purpose of a *measurement system* is to give the user a numerical value corresponding to the variable being measured. Thus a thermometer may be used to give a numerical value for the temperature of a liquid. We must, however, recognise that, for a variety of reasons, this numerical value may not actually be the true value of the variable. Thus, in the case of the thermometer, there may be errors due to the limited accuracy in the scale calibration, or reading errors due to the reading falling between two scale markings, or perhaps errors due to the insertion of a cold thermometer into a hot liquid, lowering the temperature of the liquid and so altering the temperature being measured. We thus consider a measurement system to be a block which has an input of the true value of the variable being measured and an output of the measured value of that variable (Figure 1.1).

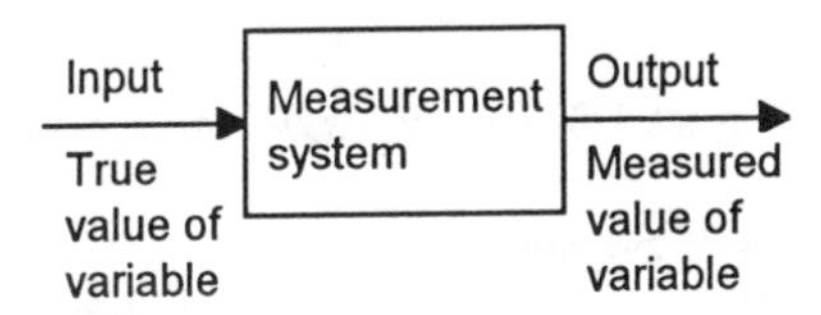

Figure 1.1 *A measurement system*

This chapter is a general consideration of measurement systems and the elements present in such systems. Chapter 2 gives a more detailed discussion of these elements and chapter 3 of measurement systems that are commonly used in engineering for temperature, pressure, flow rate, liquid level, displacement and speed measurements.

1.2 The functional elements

A measurement system consists of several elements which are used to carry out particular functions. These functional elements are:

1 *Sensor*
This is the element of the system which is effectively in contact with the process for which a variable is being measured and gives an output which depends in some way on the value of the variable. Sensors take information about the variable being measured and change it into some form which enables the rest of the measurement system to give a value to it. For example, a thermocouple is a sensor which has an input of temperature and an output of a small e.m.f. (Figure 1.2). Another example is a resistance thermometer which has an input of temperature and an output of a resistance change.

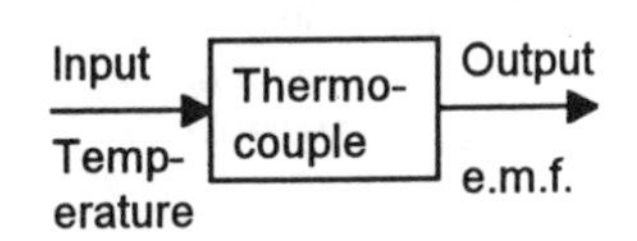

Figure 1.2 *The thermocouple sensor*

2 *Signal processor*
This element takes the output from the sensor and converts it into a form which is suitable for display or onward transmission in some control system. In the case of the thermocouple this may be an amplifier to make the e.m.f. big enough to register on a meter (Figure 1.3(a)). There often may be more than item, perhaps an element which puts the output from the sensor into a suitable condition for further processing and then an element which processes the signal so that it can be displayed. Thus in the case of the resistance thermometer there might be a signal conditioner, a Wheatstone bridge, which transforms the resistance change into a voltage change, then an amplifier to make the voltage big enough for display (Figure 1.3(b)).

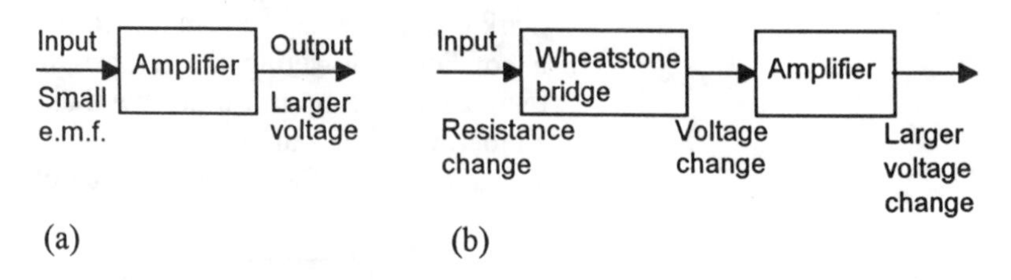

Figure 1.3 *Examples of signal processing*

3 *Data presentation*
This presents the measured value in a form which enables an observer to recognise it (Figure 1.4). This may be via a display, e.g. a pointer moving across the scale of a meter or perhaps information on a visual display unit (VDU). Alternatively, or additionally, the signal may be recorded, e.g. on the paper of a chart recorder or perhaps on magnetic disc. Alternatively, or additionally, the output from signal processing may be transmitted into other elements in a control system so that it can be compared with a required value and action initiated to control the variable.

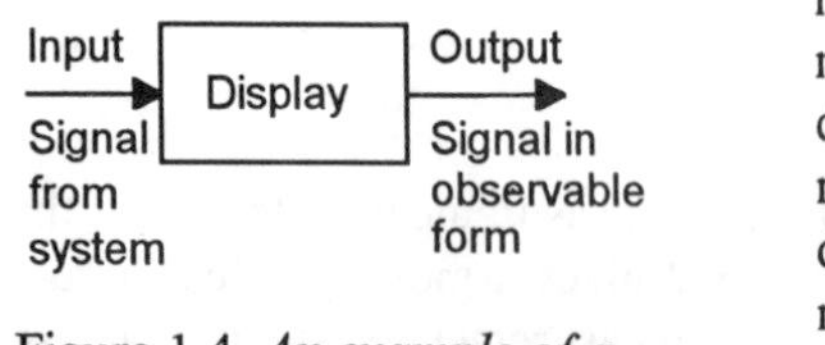

Figure 1.4 *An example of a data presentation element*

Figure 1.5 shows how these basic functional elements form a measurement system.

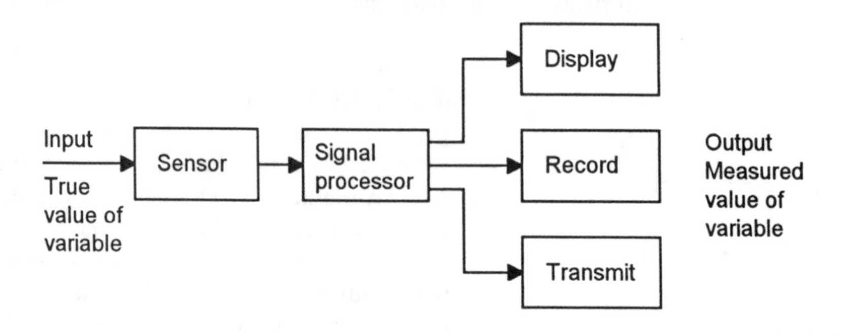

Figure 1.5 *Measurement system elements*

Example

With a resistance thermometer, element A takes the temperature signal and transforms it into resistance signal, element B transforms the resistance signal into a current signal, element C transforms the current signal into a display of a movement of a pointer across a scale. Which of these elements is (a) the sensor, (b) the signal processor, (c) the data presentation?

The sensor is element A, the signal processor element B and the data presentation element is C. The system can be represented by Figure 1.6.

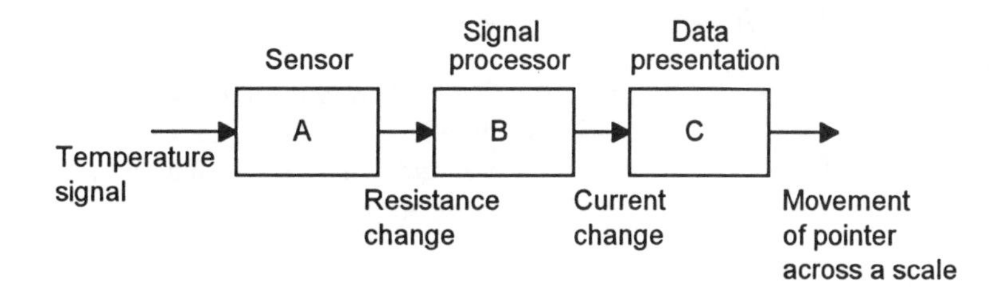

Figure 1.6 *Example*

1.3 Loading

When a cold thermometer is put in to a hot liquid to measure its temperature, the presence of the cold thermometer in the hot liquid changes the temperature of the liquid. The liquid cools and so the thermometer ends up measuring a lower temperature than that which existed before the thermometer was introduced. The act of attempting to make the measurement has modified the temperature being measured. This effect is called *loading*. If we want this modification to be small, then the thermometer should have a small heat capacity compared with that of the liquid. A small heat capacity means that very little heat is needed to change its temperature. Thus the heat taken from the liquid is minimised and so its temperature little affected.

Loading is a problem that is often encountered when measurements are being made. For example, when an ammeter is inserted into a circuit to make a measurement of the circuit current, it changes the resistance of the circuit and so changes the current being measured (Figure 1.7). The act of attempting to make such a measurement has modified the current that was being measured. If the effect of inserting the ammeter is to be as small as possible and for the ammeter to indicate the original current, the resistance of the ammeter must be very small when compared with that of the circuit.

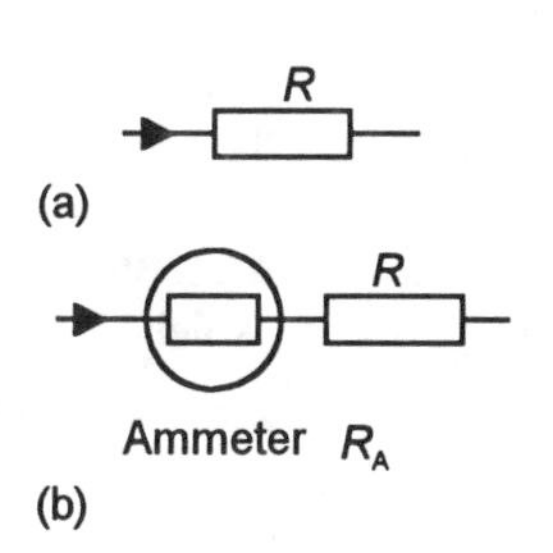

Figure 1.7 *Loading with an ammeter: (a) circuit before meter introduced, (b) extra resistance introduced by meter*

When a voltmeter is connected across a resistor to measure the voltage across it, then what we have done is connected a resistance, that of the voltmeter, in parallel with the resistance across which the voltage is to be measured. If the resistance of the voltmeter is not considerably higher than that of the resistor, the current through the resistor is markedly changed by the current passing through the meter resistance and so the voltage being measured is changed (Figure 1.8). The act of attempting to make the

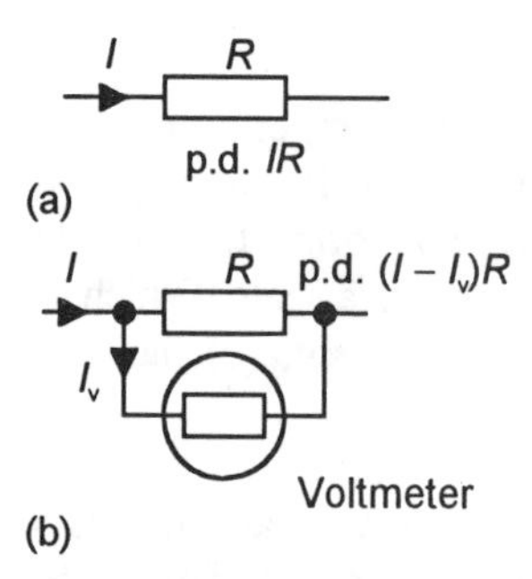

Figure 1.8 *Loading with a voltmeter: (a) before meter introduced, (b) with meter present*

measurement has modified the voltage that was being measured. If the effect of inserting the voltmeter in the circuit is to be as small as possible, the resistance of the voltmeter must be much larger than that of the resistance across which it is connected. Only then will the current bypassing the resistor and passing through the voltmeter be very small and so the voltage not significantly changed.

Example

Two voltmeters are available, one with a resistance of 1 kΩ and the other 1 MΩ. Which instrument should be selected if the indicated value is to be closest to the voltage value that existed across a 2 kΩ resistor before the voltmeter was connected across it?

The 1 MΩ voltmeter should be chosen. This is because when it is in parallel with 2 kΩ, less current will flow through it than if the 1 kΩ voltmeter had been used and so the current through the resistor will be closer to its original value.

1.4 Requirements

What is required of a measurement system? The main requirement is *fitness for purpose*. This means that if, for example, a length of a product has to be measured to a certain accuracy that the measurement system is able to be used to carry out such a measurement to the required accuracy. The *accuracy* of a measurement system is the extent to which the value it gives for a variable might be wrong. For example, a length measurement system might be quoted as having an accuracy of ±1 mm. This would mean that all the length values it gives are only guaranteed to this accuracy, e.g. for a measurement which gave a length of 120 mm the actual value could only be guaranteed to be between 119 and 121 mm. If the fitness for purpose criterion for such a measurement system is that the length can be measured to an accuracy of ±1 mm then the system is fit for that purpose. If, however, the criterion is for a system with an accuracy of ±0.5 mm then the system is not fit for that purpose.

In order to deliver the required accuracy, the measurement system must have been calibrated to give that accuracy. *Calibration* is the process of comparing the output of a measurement system against standards of known accuracy. The standards may be other measurement systems which are kept specially for calibration duties or some means of defining standard values. In many companies some instruments and items such as standard resistors and cells are kept in a company standards department and used solely for calibration purposes.

1.4.1 Quality

The British Standard BS5750 (European Standard EN 29001 and International Standard ISO 9001) is the standard which lays down the standard for a quality system. The term *quality* is used to mean that a product is one which is fit for its purpose or meets requirements, a quality system being one which makes sure that a company delivers quality products. The standard definition for quality is that it is the totality of features and characteristics of a product or service that bear on its ability to meet stated or implied needs. In everyday language the term quality tends to be used to indicate the best available. For example, a Rolls Royce might be considered to be a quality car but a small hatchback not. But in the way the term quality is used in engineering, both cars can be quality cars if they both meet the needs of those buying them, i.e. both are fit for the purpose for which they were bought. If either of the cars breaks down regularly or the paint work blisters or some other defects occur which means that the purchaser does not consider his or her needs are being met, then they are not considered quality goods.

In order to have a quality system it is necessary for a company to exercise control over its measurement systems. For example, it is not possible for a company to state that a product meets, say, a particular length specification if the measurement system used to measure the lengths does not meet the accuracy requirements of that specification. Thus a company in following the standard is expected to provide, control, calibrate and maintain inspection, measuring and test equipment suitable to demonstrate the performance of the product to the specified requirements.

The Standard lays down procedures that have to be followed when selecting, using, calibrating, controlling and maintaining measurement standards and measuring equipment. These include:

1 The company has to establish and maintain an effective system for the control and calibration of measurement standards and measuring equipment. This might involve in-company calibration or the use of a suitable calibration service.
2 All the personnel involved in the calibrating should have adequate training.
3 The calibration system used must be periodically and systematically reviewed to ensure that it continues to be effective.
4 All measurements, whether for calibration purposes or measurements of products, must take into account all the errors and uncertainties involved in the measurement process.
5 The procedures used for calibration need to be documented.
6 A separate calibration record should be kept for each measurement instrument. This record is likely to contain a description of the instrument and its reference number, the calibration date, the calibration results, how frequently the instrument is to be calibrated and probably details of the calibration procedure to be used, details of any repairs or modifications made to the instrument, and any limitations on its use.
7 The calibration should be carried out using equipment which can be traceable back to national standards.

1.4.2 Traceable standards

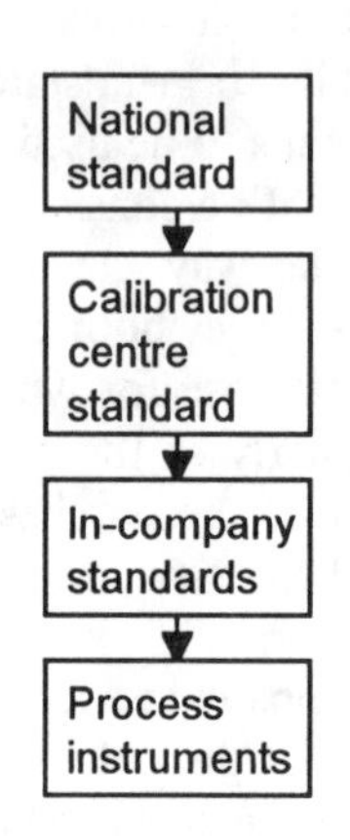

Figure 1.9 *Traceability chain*

The equipment used in the calibration of an instrument in everyday company use is likely to be *traceable* back to national standards in the following way:

1 National standards are used to calibrate standards for calibration centres.
2 Calibration centre standards are used to calibrate standards for instrument manufacturers.
3 Standardised instruments from instrument manufacturers are used to provide in-company standards.
4 In-company standards are used to calibrate process instruments.

There is a simple traceability chain from the instrument used in a process back to national standards.

The *national standards* are defined by international agreement and are maintained by national establishments, e.g. the National Physical Laboratory in Great Britain and the National Bureau of Standards in the United States. There are seven such *primary standards*, and two supplementary ones. The seven are:

1 *Mass*
The mass standard, the kilogram, is defined as being the mass of an alloy cylinder (90% platinum–10% iridium) of equal height and diameter, held at the International Bureau of Weights and Measures at Sèvres in France. Duplicates of this standard are held in other countries.

2 *Length*
The length standard, the metre, is defined as a the length of the path travelled by light in a vacuum during a time interval of duration 1/299 792 458 of a second.

3 *Time*
The time standard, the second, is defined as a time duration of 9 192 631 770 periods of oscillation of the radiation emitted by the caesium-133 atom under precisely defined conditions of resonance.

4 *Current*
The current standard, the ampere, is defined as that constant current which, if maintained in two straight parallel conductors of infinite length, of negligible circular cross-section, and placed one metre apart in a vacuum, would produce between these conductors a force equal to 2×10^{-7} N per metre of length.

5 *Temperature*
The kelvin (K) is defined so that the temperature at which liquid water, water vapour and ice are in equilibrium (known as the triple point) is 273.16 K.

6 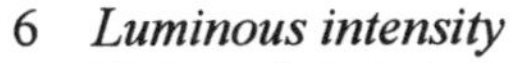
Luminous intensity
The candela is defined as the luminous intensity, in a given direction, of a specified source that emits monochromatic radiation of frequency 540×10^{12} Hz and that has a radiant intensity of 1/683 watt per unit steradian (a unit solid angle, see below).

7 *Amount of substance*
The mole is defined as the amount of a substance which contains as many elementary entities as there are atoms in 0.012 kg of the carbon 12 isotope.

The two supplementary standards are:

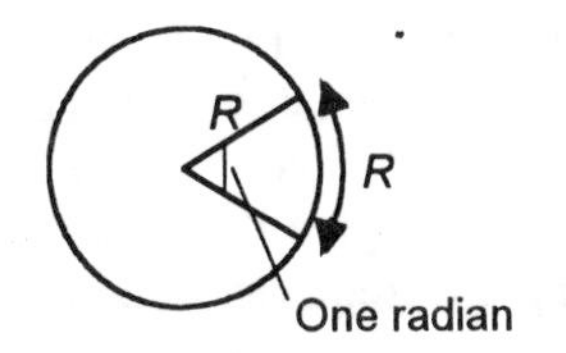

Figure 1.10 *The radian*

1 
Plane angle
The radian is the plane angle between two radii of a circle which cuts off on the circumference an arc with a length equal to the radius (Figure 1.10).

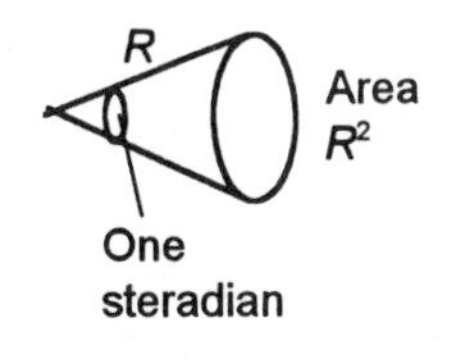

Figure 1.11 *The steradian*

2 
Solid angle
The steradian is the solid angle of a cone which, having its vertex in the centre of the sphere, cuts off an area of the surface of the sphere equal to the square of the radius (Figure 1.11).

Primary standards are used to define national standards, not only in the primary quantities but also in other quantities which can be derived from them. For example, a resistance standard of a coil of manganin wire is defined in terms of the primary quantities of length, mass, time and current. Typically these national standards in turn are used to define reference standards which can be used by national bodies for the calibration of standards which are held in calibration centres.

1.4.3 Reliability

If you toss a coin ten times you might find, for example, that it lands heads uppermost six times out of the ten. If, however, you toss the coin for a very large number of times then it is likely that it will land heads uppermost half of the times. The probability of it landing heads uppermost is said to be half. The *probability* of a particular event occurring is defined as being

$$\text{probability} = \frac{\text{number of occurrences of the event}}{\text{total number of trials}}$$

when the total number of trials is very large. The probability of the coin landing with either a heads or tails uppermost is likely to be one, since every time the coin is tossed this event will occur. A probability of 1 means a certainty that the event will take place every time. The probability of the coin landing standing on edge can be considered to be zero, since the number of occurrences of such an event is zero. The closer the probability is to 1 the more frequent an event will occur; the closer it is to zero the less frequent it will occur.

Reliability is an important requirement of a measurement system. The *reliability* of a measurement system, or element in such a system, is defined as being the probability that it will operate to an agreed level of performance, for a specified period, subject to specified environmental conditions. The agreed level of performance might be that the measurement system gives a particular accuracy. The reliability of a measurement system is likely to change with time as a result of perhaps springs slowly stretching with time, resistance values changing as a result of moisture absorption, wear on contacts and general damage due to environmental conditions. For example, just after a measurement system has been calibrated, the reliability should be 1. However, after perhaps six months the reliability might have dropped to 0.7. Thus the system cannot then be relied on to always give the required accuracy of measurement, it typically only giving the required accuracy seven times in ten measurements, seventy times in a hundred measurements.

A high reliability system will have a low failure rate. *Failure rate* is the number of times during some period of time that the system fails to meet the required level of performance, i.e.:

$$\text{Failure rate} = \frac{\text{number of failures}}{\text{number of systems observed} \times \text{time observed}}$$

A failure rate of 0.4 per year means that in one year, if ten systems are observed, that 4 will fail to meet the required level of performance. If 100 systems are observed, 40 will fail to meet the required level of performance. Failure rate is affected by environmental conditions. For example, the failure rate for a temperature measurement system used in hot, dusty, humid, corrosive conditions might be 1.2 per year, while for the same system used in dry, cool, non-corrosive environment it might be 0.3 per year.

With a measurement system consisting of a number of elements, failure occurs when just one of the elements fails to reach the required performance. Thus in a system for the measurement of the temperature of a fluid in some plant we might have a thermocouple, an amplifier and a meter. The failure rate is likely to be highest for the thermocouple since that is the element in contact with the fluid while the other elements are likely to be a the controlled atmosphere of a control room. The reliability of the system might thus be markedly improved by choosing materials for the thermocouple which resist attack by the fluid. Thus it might be in a stainless steel sheath to prevent fluid coming into direct contact with the thermocouple wires.

Example

The failure rate for a pressure measurement system used in factory A is found to be 1.0 per year while the system used in factory B is 3.0 per year. Which factory has the most reliable pressure measurement system?

The higher the reliability the lower the failure rate. Thus factory A has the more reliable system. The failure rate of 1.0 per year means that if 100 instruments are checked over a period of a year, 100 failures will be found, i.e. on average each instrument is failing once. The failure rate of 3.0 means that if 100 instruments are checked over a period of a year, 300 failures will be found, i.e. instruments are failing more than once in the year.

1.4.3 Repeatability

Suppose you use a particular steel rule to measure the length of a constant length rod and repeat the measurement over a number of days. The results obtained might be:

20.1 mm, 20.2 mm, 20.1 mm, 20.0 mm, 20.1 mm, etc.

The results of the measurement give values scattered about some value. The term *repeatability* is used for the ability of a measurement system to give the same value for repeated measurements of the same value of a variable. The most common cause of lack of repeatability are random fluctuations in the environment, e.g. changes in temperature and humidity.

1.5 Testing

Testing a measurement system installation can be considered to fall into three stages:

1. *Pre-installation testing*
   This is the testing of each instrument and element for correct calibration and operation prior to it being installed as part of a measurement system.

2. *Cabling and piping testing*
   Cables and/or piping will be used to connect together the elements of measurement systems. The display might, for example, be in a control room. All the instrument cables should be checked for continuity and insulation resistance prior to the connection of any instruments or elements of the system. When the system involves pneumatic lines the testing involves blowing through with clear, dry, air prior to connection and pressure testing to ensure they are leak free.

3. *Precommissioning*
   This involves testing that the measurement system installation is complete, all the instrument and other components are in full operational order when interconnected and all control room panels or displays function.

Problems

Questions 1 to 7 have four answer options: A. B, C and D. Choose the correct answer from the answer options.

1 Decide whether each of these statements is True (T) or False (F).

Sensors in a measurement system have:
(i) An input of the variable being measured.
(ii) An output of a signal in a form suitable for further processing in the measurement system.

Which option BEST describes the two statements?

A (i) T (ii) T
B (i) T (ii) F
C (i) F (ii) T
D (i) F (ii) F

2 The following lists the types of signals that occur in sequence at the various stages in a particular measurement system:

(i) Temperature
(ii) Voltage
(iii) Bigger voltage
(iv) Movement of pointer across a scale

The signal processor is the functional element in the measurement system that changes the signal from:

A (i) to (ii)
B (ii) to (iii)
C (iii) to (iv)
D (ii) to (iv)

3 Decide whether each of these statements is True (T) or False (F).

The discrepancy between the measured value of the current in a electrical circuit and the value before the measurement system, an ammeter, was inserted in the circuit is bigger the larger:
(i) The resistance of the meter.
(ii) The resistance of the circuit.

Which option BEST describes the two statements?

A (i) T (ii) T
B (i) T (ii) F
C (i) F (ii) T
D (i) F (ii) F

4 Decide whether each of these statements is True (T) or False (F).

A highly reliable measurement system is one where there is a high chance that the system will:
(i) Require frequent calibration.
(ii) Operate to the specified level of performance.

Which option BEST describes the two statements?

A (i) T (ii) T
B (i) T (ii) F
C (i) F (ii) T
D (i) F (ii) F

5 Decide whether each of these statements is True (T) or False (F).

A measurement system which has a lack of repeatability is one where there could be:
(i) Random fluctuations in the values given by repeated measurements of the same variable.
(ii) Fluctuations in the values obtained by repeating measurements over a number of samples.

Which option BEST describes the two statements?

A (i) T (ii) T
B (i) T (ii) F
C (i) F (ii) T
D (i) F (ii) F

6 Decide whether each of these statements is True (T) or False (F).

For a measurement system to be of the right quality and so fit for the required purpose it must have:
(i) The highest possible accuracy.
(ii) Been calibrated directly against national standards.

Which option BEST describes the two statements?

A (i) T (ii) T
B (i) T (ii) F
C (i) F (ii) T
D (i) F (ii) F

7 Decide whether each of these statements is True (T) or False (F).

For a measurement system to be of the right quality and so fit for the required purpose it must have:
(i) The accuracy needed for the measurement in question.
(ii) Been calibrated using a calibration system that is periodically and systematically reviewed to ensure that it continues to be effective.

Which option BEST describes the two statements?

A (i) T (ii) T
B (i) T (ii) F
C (i) F (ii) T
D (i) F (ii) F

8 List and explain the functional elements of a measurement system.

9 Explain the terms (a) reliability and (b) repeatability when applied to a measurement system.

10 Explain what is meant by calibration standards having to be traceable to national standards.

11 Explain what is meant by 'fitness for purpose' when applied to a measurement system.

12 The reliability of a measurement system is said to be 0.6. What does this mean?

13 The measurement instruments used in the tool room of a company are found to have a failure rate of 0.01 per year. What does this mean?

2 Functional elements

2.1 Functional elements

This chapter presents a summary of sensors, signal processors and data presentation elements commonly used in engineering. The term *sensor* is used for an element which produces a signal relating to the quantity being measured. The term *signal processor* is used for the element that takes the output from the sensor and converts it into a form which is suitable for data presentation. *Data presentation* is where the data is displayed, recorded or transmitted to some control system.

The term *transducer* is often used in relation to measurement systems. Transducers are defined as an element that converts a change in some physical variable into a related change in some other physical variable. It is generally used for an element that converts a change in some physical variable into a change in an electrical signal. Thus sensors can be transducers. However, a measurement system may use transducers, in addition to the sensor, in other parts of the system to convert signals in one form to another form.

2.1.1 Performance terms

The following are some of the more common terms used to define the performance of measurement systems, and functional elements.

Accuracy and error

Accuracy is the extent to which the value indicated by a measurement system or element might be wrong. For example, a thermometer may have an accuracy of ±0.1°C. Accuracy is often expressed as a percentage of the full range output or full-scale deflection (f.s.d). For example, a system might have an accuracy of ±1% of f.s.d. If the full-scale deflection is, say, 10 A, then the accuracy is ±0.1 A. The accuracy is a summation of all the possible errors that are likely to occur, as well as the accuracy to which the system or element has been calibrated.

The term *error* is used for the difference between the result of the measurement and the true value of the quantity being measured, i.e.

$$\text{Error} = \text{measured value} - \text{true value}$$

Thus if the measured value is 10.1 when the true value is 10.0, the error is +0.1. If the measured value is 9.9 when the true value is 10.0, the error is −0.1.

The term *hysteresis error* (Figure 2.1) is used for the difference in outputs given when different outputs are given from the same value of quantity being measured according to whether that value has been reached by a continuously increasing change or a continuously decreasing change.

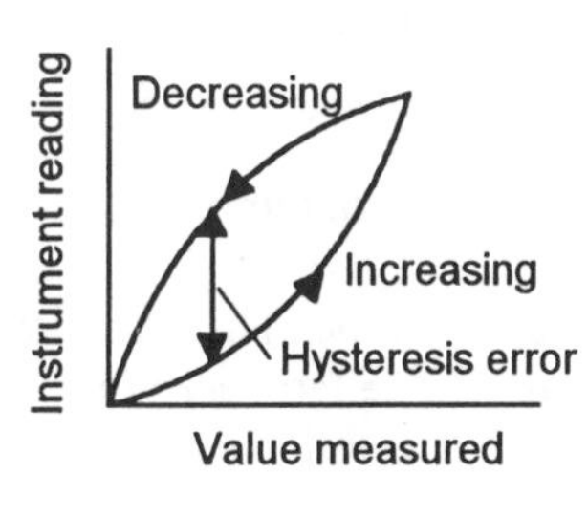

Figure 2.1 *Hysteresis error*

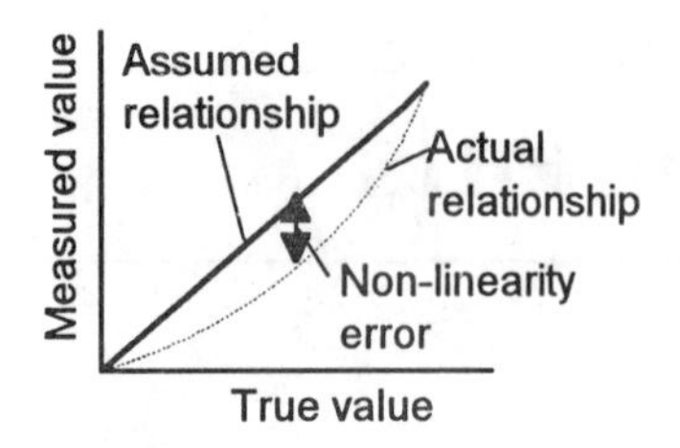

Figure 2.2 *Non-linearity error*

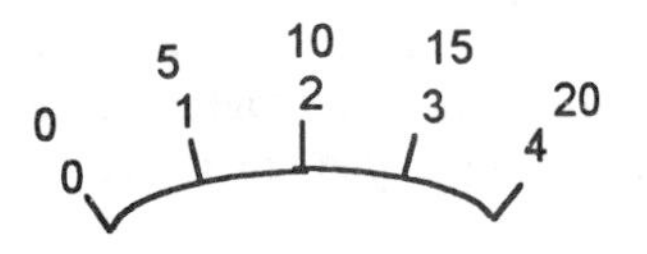

Figure 2.3 *Multi-range meter*

The term *non-linearity error* (Figure 2.2) is used for the error that occurs as a result of assuming a linear relationship between the input and output over the working range, i.e. a graph of output plotted against input is assumed to give a straight line. Few systems or elements, however, have a truly linear relationship and thus errors occur as a result of the assumption of linearity.

Range

The range of an element or system is the limits between which the input can vary. For example, a resistance thermometer sensor might be quoted as having a range of –200 to +800°C. The meter shown in Figure 2.3 has the dual ranges 0 to 4 and 0 to 20.

The term *dead band* or *dead space* is used if there is a range of input values for which there is no output. For example, bearing friction in a flow meter using a rotor might mean that there is no output until the input has reached a particular flow rate threshold.

Repeatability

The repeatability of a system or element is its ability to give the same output for repeated applications of the same input value, without the system or element being disconnected from its input or any change in the environment in which the test is carried out (see section 1.4.3). The resulting error is usually expressed as a percentage of the full range output. For example, a pressure sensor might be quoted as having a repeatability of ±0.1% of full range. Thus with a range of 20 kPa this would be an error of ±20 Pa.

Reproducibility

The reproducibility of a system or element is its ability to give the same output when used with a constant input with the system or element being disconnected from its input and then reinstalled. The resulting error is usually expressed as a percentage of the full range output.

Sensitivity

The sensitivity indicates how much output you get per unit input, i.e. the ratio ouput/input. For example, a thermocouple might have a sensitivity of 20 μV/°C. This term is also frequently used to indicate the sensitivity to inputs other than that being measured, i.e. environmental changes. For example, the sensitivity of a system or element might be quoted to changes in temperature or perhaps fluctuations in the mains voltage supply. Thus a pressure measurement sensor might be quoted as having a temperature sensitivity of ±0.1% of the reading per °C change in temperature.

Stability

The stability of a transducer is its ability to give the same output when used to measure a constant input over a period of time. The term *drift* is often used to describe the change in output that occurs over time. The drift may be expressed as a percentage of the full range output. The term *zero drift* is used for the changes that occur in output when there is zero input.

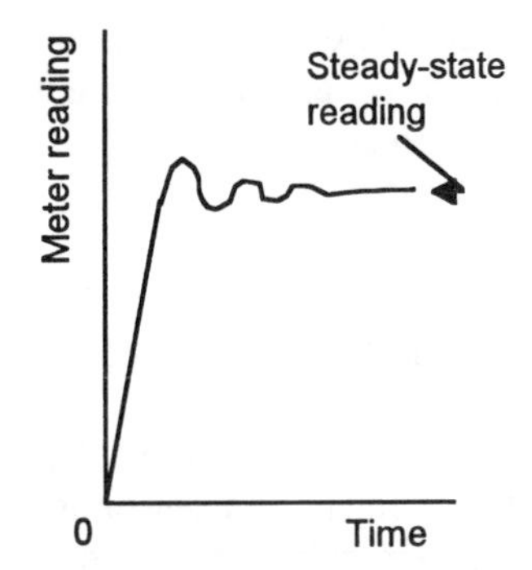

Figure 2.4 *Oscillations of a meter reading*

The terms given above refer to what can be termed the *static characteristics*. These are the values given when steady-state conditions occur, i.e. the values given when the system or element has settled down after having received some input. The *dynamic characteristics* refer to the behaviour between the time that the input value changes and the time that the value given by the system or element settles down to the steady-state value. For example, Figure 2.4 shows how the reading of an ammeter might change when the current is switched on. The meter pointer oscillates before settling down to give the steady-state reading. The following are terms commonly used for dynamic characteristics.

Response time
This is the time which elapses after the input to a system or element is abruptly increased from zero to a constant value up to the point at which the system or element gives an output corresponding to some specified percentage, e.g. 95%, of the value of the input.

Rise time
This is the time taken for the output to rise to some specified percentage of the steady-state output. Often the rise time refers to the time taken for the output to rise from 10% of the steady-state value to 90 or 95% of the steady-state value.

Settling time
This is the time taken for the output to settle to within some percentage, e.g. 2%, of the steady state value.

Example

A pressure measurement system (a diaphragm sensor giving a capacitance change with output processed by a bridge circuit and displayed on a digital display) is stated as having the following characteristics. Explain the significance of the terms:

Range: 0 to 125 kPa and 0 to 2500 kPa
Accuracy: ±1% of the displayed reading
Temperature sensitivity: ±0.1% of the reading per °C

The range indicates that the system can be used to measure pressures from 0 to 125 kPa or 0 to 2500 kPa. The accuracy is expressed as a percentage of the displayed reading, thus if the instrument indicates a pressure of, say, 100 kPa then the error will be ±1 kPa. The temperature sensitivity indicates that if the temperature changes by 1°C that displayed reading will be in error by ±0.1% of the value. Thus for a pressure of, say, 100 kPa the error will be ±0.1 kPa for a 1°C temperature change.

2.2 Sensors

The following are examples of sensors grouped according to what they are being used to measure. The measurements considered are displacement, speed, pressure, fluid flow, liquid level and temperature.

2.2.1 Displacement

A displacement sensor is here considered to be one that can be used to:

1 Measure a linear displacement, i.e. a change in linear position. This might, for example, be the change in linear displacement of a sensor as a result of a change in the thickness of sheet metal emerging from rollers.
2 Measure an angular displacement, i.e. a change in angular position. This might, for example, be the change in angular displacement of a drive shaft.
3 Measure the position, linear, or angular, of some object.
4 Detect motion or the presence of some object. This might be as part of an alarm or automatic light system, whereby an alarm is sounded or a light switched on when there is some movement of an object within the 'view' of the sensor.

Displacement sensors fall into two groups: those that make direct contact with the object being monitored, by spring loading or mechanical connection with the object, and those which are non-contacting. For those linear displacement methods involving contact, there is usually a sensing shaft which is in direct contact with the object being monitored, the displacement of this shaft is then being monitored by a sensor. This shaft movement may be used to cause changes in electrical voltage, resistance, capacitance, or mutual inductance. For angular displacement methods involving mechanical connection, the rotation of a shaft might directly drive, through gears, the rotation of the sensor element, this perhaps generating an e.m.f. Non-contacting displacement sensors might consist of a beam of infrared light being broken by the presence of the object being monitored, the sensor then giving a voltage signal indicating the breaking of the beam, or perhaps the beam being reflected from the object being monitored, the sensor giving a voltage indicating that the reflected beam has been detected.

The following are examples of displacement sensors.

Potentiometer

A potentiometer consists of a resistance element with a sliding contact which can be moved over the length of the element and connected as shown in Figure 2.5. With a constant supply voltage V_s, the output voltage V_o between terminals 1 and 2 is a fraction of the input voltage, the fraction depending on the ratio of the resistance R_{12} between terminals 1 and 2 compared with the total resistance R of the entire length of the track across which the supply voltage is connected. Thus $V_o/V_s = R_{12}/R$. If the track has a constant resistance per unit length, the output is proportional to the displacement of the slider from position 1. A rotary potentiometer consists of a coil of wire wrapped round into a circular track or a circular film of conductive plastic over which a rotatable sliding contact can be rotated,

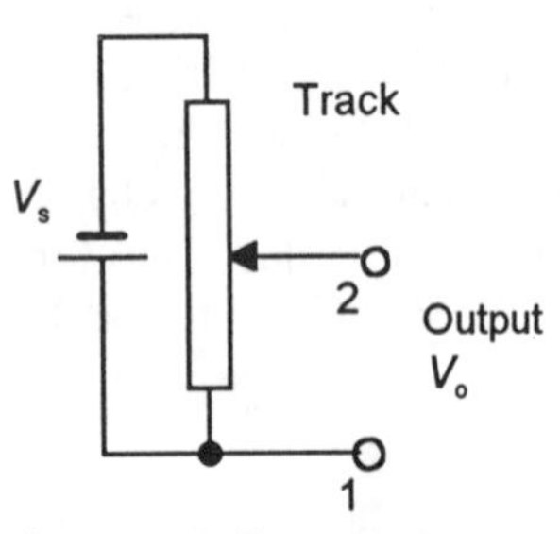

Figure 2.5 *Potentiometer*

hence an angular displacement can be converted into a potential difference. Linear tracks can be used for linear displacements. With a wire wound track the output voltage does not continuously vary as the slider is moved over the track but goes in small jumps as the slider moves from one turn of wire to the next. This problem does not occur with a conductive plastic track. Errors due to non-linearity of the track for wire tracks tend to range from less than 0.1% to about 1% of the full range output and for conductive plastics are of the order of 0.05%. The track resistance for wire-wound potentiometers tends to range from about 20 Ω to 200 kΩ and for conductive plastic from about 500 Ω to 80 kΩ. The conductive plastic has a higher temperature coefficient of resistance than the wire and so temperature changes have a greater effect on accuracy.

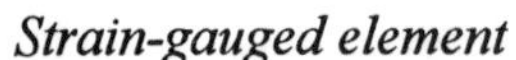

Strain-gauged element

Figure 2.6 shows the basic form of an electrical resistance strain gauge. Strain gauges consist of a flat length of metal wire, metal foil strip, or a strip of semi-conductor material which can be stuck onto surfaces like a postage stamp. When the wire, foil, strip or semiconductor is stretched, its resistance R changes. The fractional change in resistance $\Delta R/R$ is proportional to the strain ε, i.e.:

Lead Lead Paper backing Gauge wire

Figure 2.6 *Strain gauge*

$$\frac{\Delta R}{R} = G\varepsilon$$

where G, the constant of proportionality, is termed the gauge factor. Metal strain gauges typically have gauge factors of the order of 2.0. When such a strain gauge is stretched its resistance increases, when compressed its resistance decreases. Strain is (change in length/original length) and so the resistance change of a strain gauge is a measurement of the change in length of the gauge and hence the surface to which the strain gauge is attached. Thus a displacement sensor might be constructed by attaching strain gauges to a cantilever (Figure 2.7), the free end of the cantilever being moved as a result of the linear displacement being monitored. When the cantilever is bent, the electrical resistance strain gauges mounted on the element are strained and so give a resistance change which can be monitored and which is a measure of the displacement. With strain gauges mounted as shown in Figure 2.7, when the cantilever is deflected downwards the gauge on the upper surface is stretched and the gauge on the lower surface compressed. Thus the gauge on the upper surface increases in resistance while that on the lower surface decreases. Typically, this type of sensor is used for linear displacements of the order of 1 mm to 30 mm, having a non-linearity error of about ± 1% of full range.

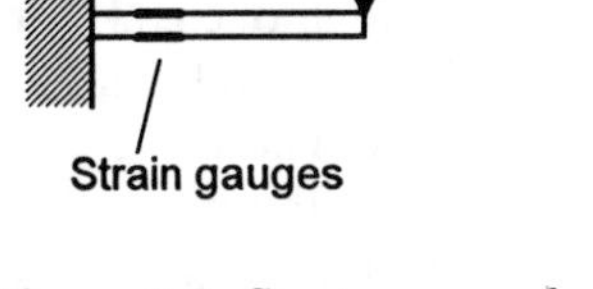

Figure 2.7 *Strain-gauged cantilever*

Capacitive element

The capacitance C of a parallel plate capacitor (Figure 2.8) is given by:

$$C = \frac{\varepsilon_r \varepsilon_0 A}{d}$$

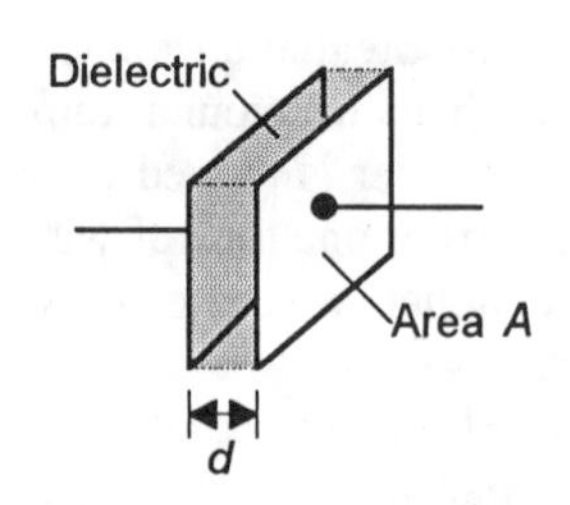

Figure 2.8 *Parallel plate capacitor*

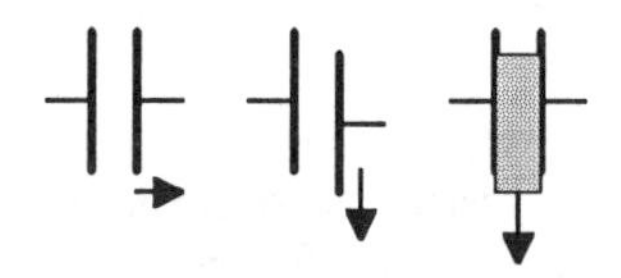

Figure 2.9 *Capacitor sensors*

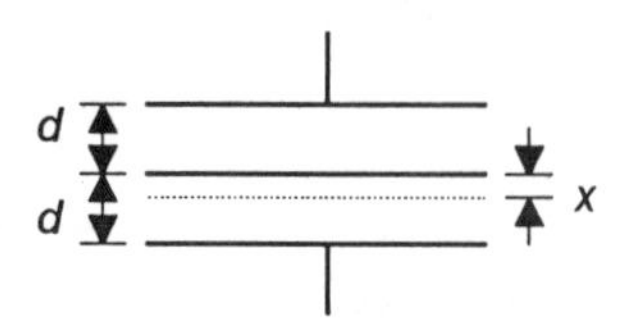

Figure 2.10 *Push-pull displacement sensor*

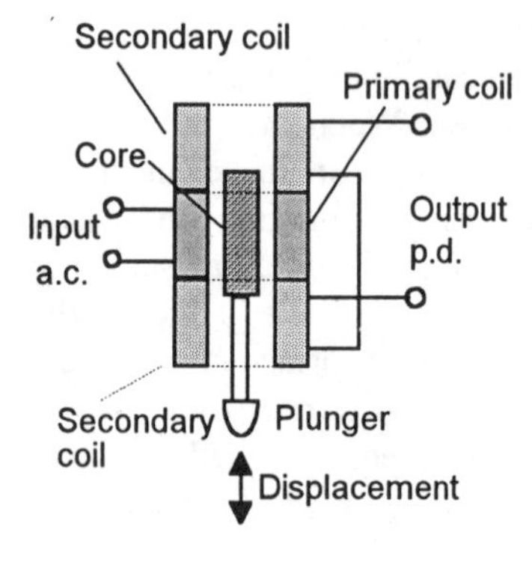

Figure 2.11 *LVDT*

where ε_r is the relative permittivity of the dielectric between the plates, ε_0 a constant called the permittivity of free space, A the area of overlap between the two plates and d the plate separation. The capacitance will change if the plate separation d changes, the area A of overlap of the plates changes, or a slab of dielectric is moved into or out of the plates, so varying the effective value of ε_r (Figure 2.9). All these methods can be used to give linear displacement sensors.

One form that is often used is shown in Figure 2.10 and is referred to as a *push-pull* displacement sensor. It consists of two capacitors; between the movable central plate and the upper plate and between the central movable plate and the lower plate. The displacement x moves the central plate between the two other plates. Thus when the central plate moves upwards it decreases the plate separation of the upper capacitor and increases the separation of the lower capacitor. Thus the capacitance of the upper capacitor is increased and that of the lower capacitor decreased. When the two capacitors are incorporated in opposite arms of an alternating current bridge, the output voltage from the bridge is proportional to the displacement. Such a sensor has good long-term stability and is typically used for monitoring displacements from a few millimetres to hundreds of millimetres. Non-linearity and hysteresis errors are about ± 0.01% of full range.

Linear variable differential transformer

The linear variable differential transformer, generally referred to by the abbreviation LVDT, is a transformer with a primary coil and two secondary coils. Figure 2.11 shows the arrangement, there being three coils symmetrically spaced along an insulated tube. The central coil is the primary coil and the other two are identical secondary coils which are connected in series in such a way that their outputs oppose each other. A magnetic core is moved through the central tube as a result of the displacement being monitored. When there is an alternating voltage input to the primary coil, alternating e.m.f.s are induced in the secondary coils. With the magnetic core in a central position, the amount of magnetic material in each of the secondary coils is the same and so the e.m.f.s induced in each coil are the same. Since they are so connected that their outputs oppose each other, the net result is zero output. However, when the core is displaced from the central position there is a greater amount of magnetic core in one coil than the other. The result is that a greater e.m.f. is induced in one coil than the other and then there is a net output from the two coils. The bigger the displacement the more of the core there is in one coil than the other, thus the difference between the two e.m.f.s increases the greater the displacement of the core. Typically, LVDTs have operating ranges from about ±2 mm to ±400 mm. Non-linearity errors are typically about ±0.25%. LVDTs are very widely used for monitoring displacements.

Optical encoders

An encoder is a device that provides a digital output as a result of an angular or linear displacement. Position encoders can be grouped into two categories: incremental encoders, which detect changes in displacement from some datum position, and absolute encoders, which give the actual position. Figure 2.12 shows the basic form of an *incremental encoder* for the measurement of angular displacement of a shaft. It consists of a disc which rotates along with the shaft. In the form shown, the rotatable disc has a number of windows through which a beam of light can pass and be detected by a suitable light sensor. When the shaft rotates and disc rotates, a pulsed output is produced by the sensor with the number of pulses being proportional to the angle through which the disc rotates. The angular displacement of the disc, and hence the shaft rotating it, can thus be determined by the number of pulses produced in the angular displacement from some datum position. Typically the number of windows on the disc varies from 60 to over a thousand with multi-tracks having slightly offset slots in each track. With 60 slots occurring with 1 revolution then, since 1 revolution is a rotation of 360°, the minimum angular displacement, i.e. the resolution, that can be detected is 360/60 = 6°. The resolution thus typically varies from about 6° to 0.3° or better.

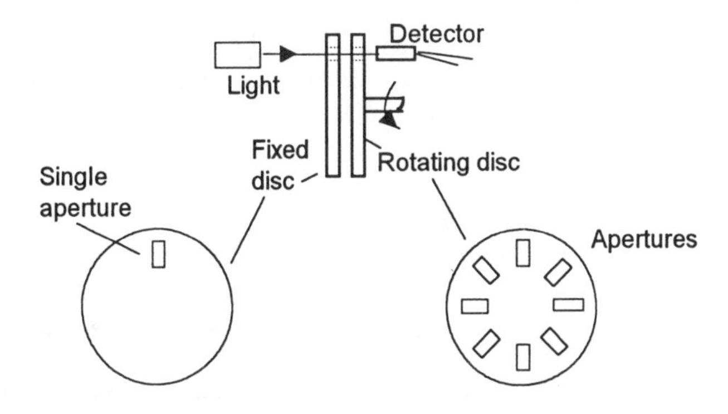

Figure 2.12 *Optical incremental encoder*

With the incremental encoder, the number of pulses counted gives the angular displacement, a displacement of, say, 50° giving the same number of pulses whatever angular position the shaft starts its rotation from. However, the absolute encoder gives an output in the form of a binary number of several digits, each such number representing a particular angular position. Figure 2.13 shows the basic form of an *absolute encoder* for the measurement of angular position. With the one shown in the figure, the rotating disc has four concentric circles of slots and four sensors to detect the light pulses. The slots are arranged in such a way that the sequential output from the sensors is a number in the binary code, each such number corresponding to a particular angular position. A number of forms of binary code are used. Typical encoders tend to have up to 10 or 12 tracks. The number of bits in the binary number will be equal to the number of tracks. Thus with 10 tracks there will be 10 bits and so the number of positions that can be detected is 2^{10}, i.e. 1024, a resolution of 360/1024 = 0.35°.

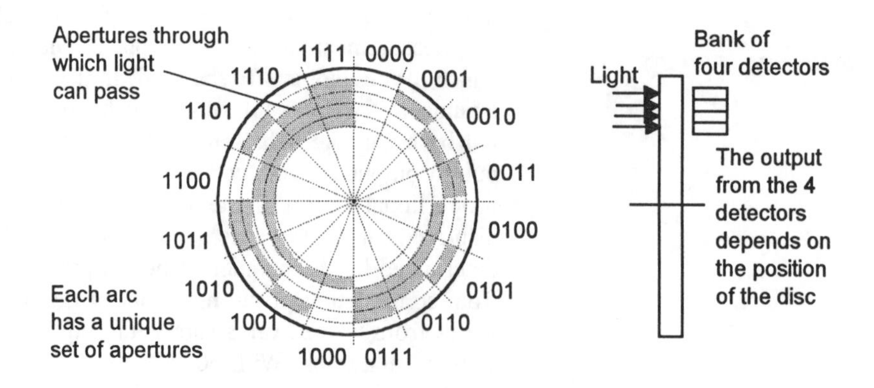

Figure 2.13 *The rotating wheel of the absolute encoder*

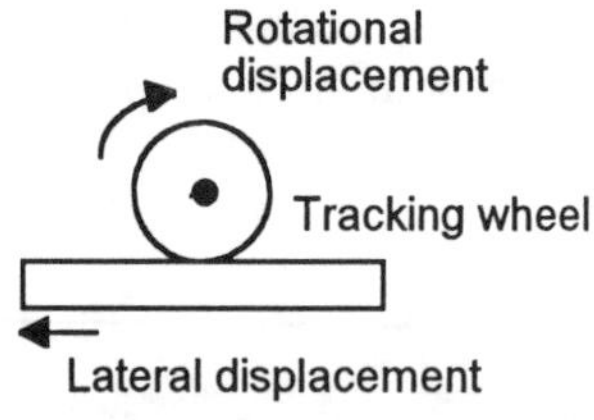

Figure 2.14 *Tracking wheel*

The incremental encoder and the absolute encoder can be used with linear displacements if the linear displacement is first converted to a rotary motion by means of a tracking wheel (Figure 2.14).

Optical sensors

There are a variety of optical sensors that can be used to detect displacement. The detection may be based on the interruption of a beam of radiation, as illustrated in the encoder shown in Figures 2.12 and 2.13, or by detecting the reflected beam from a surface. Figure 2.15 illustrates a proximity sensor based on reflection. A photodiode emits infrared radiation which is reflected by the object. Reflected radiation is then detected by a phototransistor. In the absence of the object there is no detected reflected radiation; when the object is in the proximity, there is.

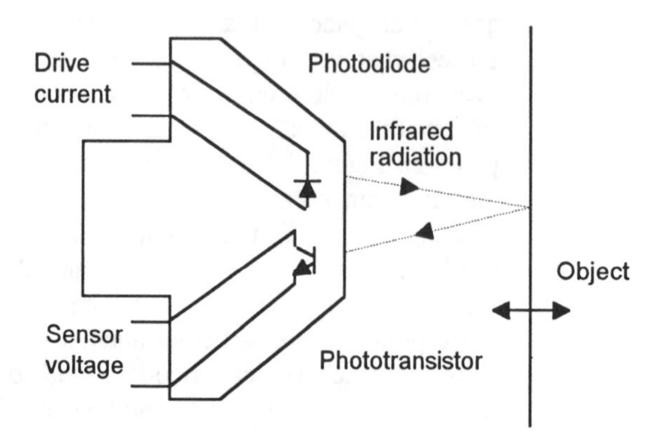

Figure 2.15 *Proximity sensor*

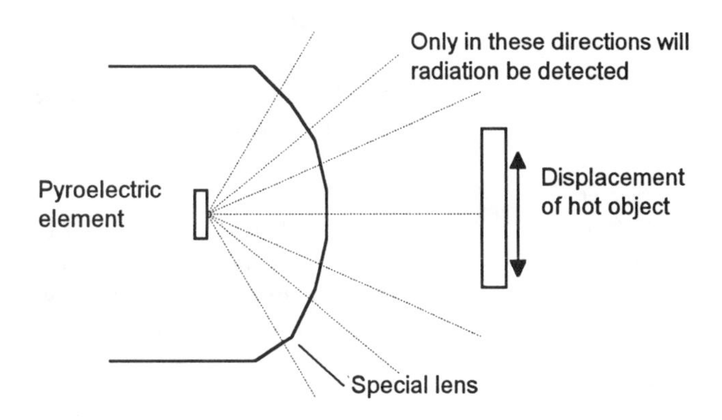

Figure 2.16 *Pyroelectric sensor*

Another form of optical sensor is the pyroelectric sensor. Such sensors give a voltage signal when the infrared radiation falling on them changes, no signal being given for constant radiation levels. Lithium tantulate is a widely used pyroelectric material. Figure 2.16 shows an example of such a sensor. Such sensors can be used with burglar alarms or for the automatic switching on of a light when someone walks up the drive to a house. A special lens is put in front of the detector. When a object which emits infra-red radiation is in front of the detector, the radiation is focused by the lens onto the detector. But only for beams of radiation in certain directions will a focused beam fall on the detector and give a signal. Thus when the object moves then the focused beam of radiation is effectively switched on and off as the object cuts across the lines at which its radiation will be detected. Thus the pyroelectric detector gives a voltage output related to the changes in the signal.

Switches

There are many situations where a sensor is required to detect the presence of some object. For example, the presence of a work piece on an assembly line might need to be detected. The sensor used in such situations can be a push button switch. Figure 2.17 illustrates such an application. Switches are used for such applications as where a work piece closes the switch by pushing against it when it reaches the correct position on a work table, such a switch being referred to as a limit switch. The switch might then be used to switch on a machine tool to carry out some operation on the work piece. Another example is a light being required to come on when a door is opened, as in a refrigerator. The action of opening the door can be made to close the contacts in a switch and trigger an electrical circuit to switch on the lamp.

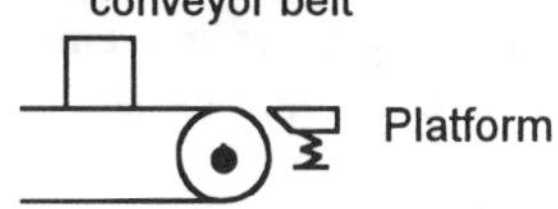

Figure 2.17 *Using a switch as a position detector*

2.2.2 Speed

The following are examples of sensors that can be used to monitor linear and angular speeds.

Optical methods
Linear speeds can be measured by determining the time between when the moving object breaks one beam of radiation and when it breaks a second beam some measured distance away (Figure 2.18). Breaking the first beam can be used to start an electronic clock and breaking the second beam to stop the clock.

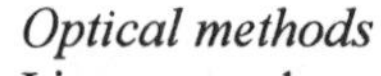

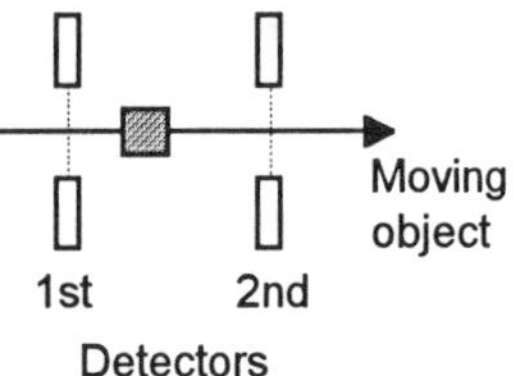

Figure 2.18 *Measurement of linear speed*

Incremental encoder
The incremental encoder described above in section 2.2.1 (Figure 2.12) can be used for a measurement of angular speed or a rotating shaft, the number of pulses produced per second being counted.

Tachogenerator
The basic tachogenerator consists of a coil mounted in a magnetic field (Figure 2.19). When the coil rotates electromagnetic induction results in an alternating e.m.f. being induced in the coil. The faster the coil rotates the greater the size of the alternating e.m.f. Thus the size of the alternating e.m.f. is a measure of the angular speed. Typically such a sensor can be used up to 10 000 revs per minute and has a non-linearity error of about ±1% of the full range.

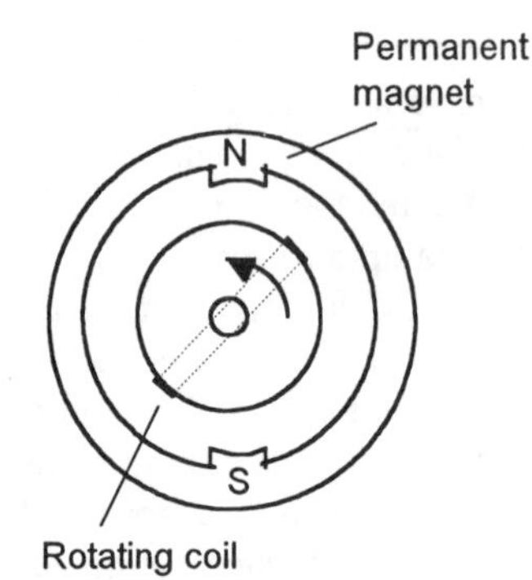

Figure 2.19 *The tacho-generator principle*

2.2.3 Fluid pressure

Many of the devices used to monitor fluid pressure in industrial processes involve the monitoring of the elastic deformation of diaphragms, bellows and tubes. The following are some common examples of such sensors.

Diaphragm sensor
The movement of the centre of a circular diaphragm as a result of a pressure difference between its two sides is the basis of a pressure gauge (Figure 2.20). For the measurement of the absolute pressure, the opposite side of the diaphragm is a vacuum, for the measurement of pressure difference the pressures are connected to each side of the diaphragm, for the gauge pressure, i.e. the pressure relative to the atmospheric pressure, the opposite side of the diaphragm is open to the atmosphere. The amount of movement with a plane diaphragm is fairly limited; greater movement can, however, be produced with a diaphragm with corrugations (Figure 2.21).

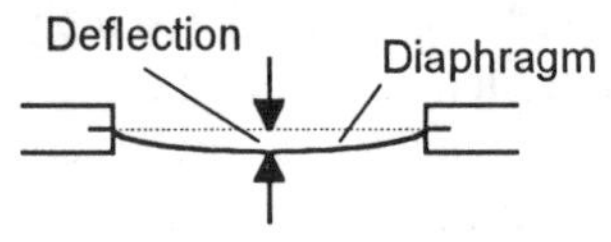

Figure 2.20 *Diaphragm sensor*

Figure 2.21 *Diaphragm with corrugations*

The movement of the centre of a diaphragm can be monitored by some form of displacement sensor. Figure 2.22(a) shows the form that might be taken when strain gauges are used to monitor the displacement, the strain gauges monitoring the bending of a cantilever as a result of the diaphragm movement. Typically such sensors are used for pressures over the range 100 kPa to 100 MPa, with an accuracy up to about ±0.1%. Figure 2.22(b) shows the form that might be taken by a capacitance diaphragm pressure gauge. The diaphragm forms one plate of a capacitor, the other plate being fixed. Displacement of the diaphragm results in changes in capacitance. The range of capacitance diaphragm pressure gauges is about 1 kPa to 200 kPa with an accuracy of about ±0.1%.

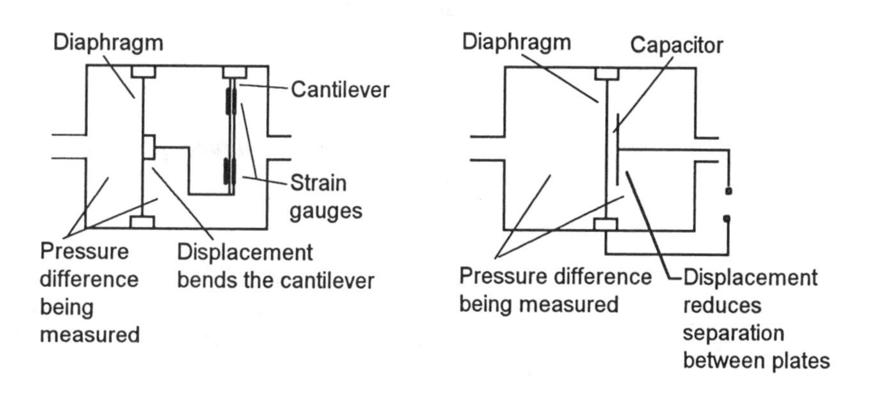

Figure 2.22 *Basic forms of: (a) a strain gauge diaphragm gauge, (b) a capacitance diaphragm gauge*

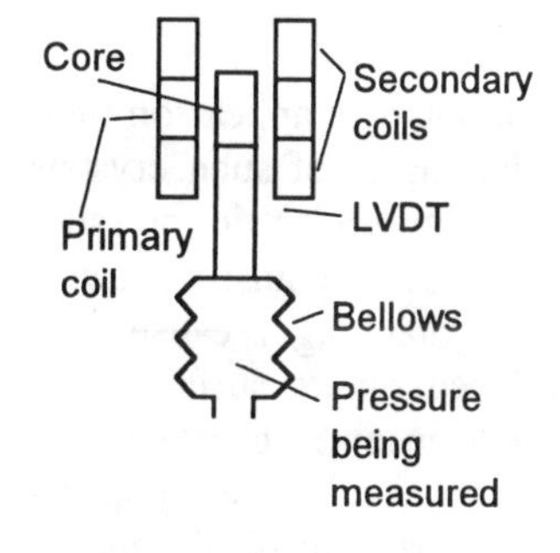

Figure 2.23 *LVDT bellows pressure gauge*

Bellows sensor

When the pressure inside bellows increases, relative to the pressure outside the bellows, then the bellows increase in length. A displacement sensor can then be used to monitor the movement of bellows and hence give a measure of the pressure. Figure 2.23 shows how a bellows can be combined with a LVDT to give a pressure sensor with an electrical output. Typically such an arrangement can be used for pressure differences up to a few hundred kPa with an accuracy of about ±1%.

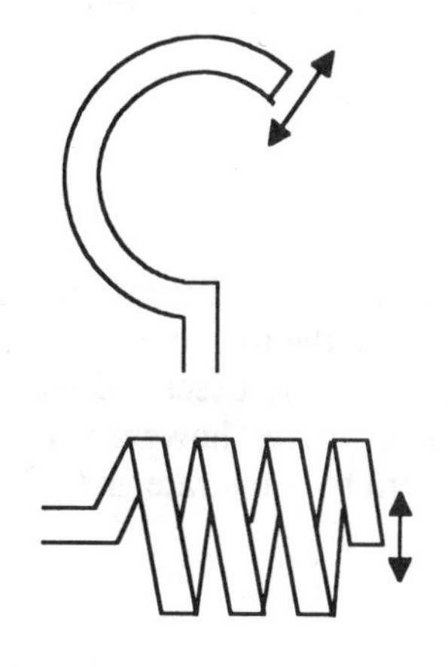

Figure 2.24 *Forms of Bourdon tubes*

Bourdon tube

The Bourdon tube is an almost rectangular or elliptical cross-section tube made from materials such as stainless steel or phosphor bronze. With a C-shaped tube (Figure 2.24(a)), when the pressure inside the tube increases the closed end of the C opens out, thus the displacement of the closed end becomes a measure of the pressure. Another form uses a helical-shaped tube (Figure 2.24(b)). When the pressure inside the tube increases, the closed end of the tube rotates and thus the rotation becomes a measure of the pressure.

Figure 2.25(a) illustrates how a C-shaped Bourdon tube can be used to rotate, via gearing, a shaft and cause a pointer to move across a scale. Such instruments are robust and typically used for pressures in the range 10 kPa to 100 MPa with an accuracy of about ±1% of full scale. Figure 2.25(b) shows how a helical-shaped Bourdon tube can be used to move the slider of

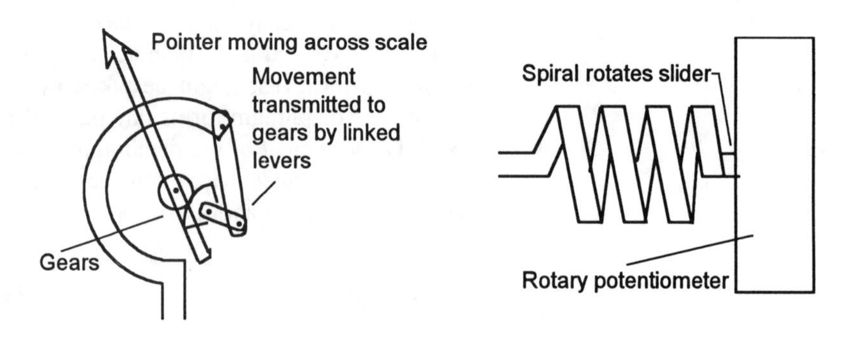

Figure 2.25 *Bourdon tube instruments: (a) geared form, (b) potentiometer form*

a potentiometer and so give an electrical output related to the pressure. Helical tubes are more expensive but have greater sensitivity. Typically they are used for pressures up to about 50 MPa with an accuracy of about ±1% of full range.

Piezoelectric sensor

When certain crystals are stretched or compressed, charges appear on their surfaces. This effect is called *piezo-electricity*. Examples of such crystals are quartz, tourmaline, and zirconate-titanate. A piezoelectric pressure gauge consists essentially of a diaphragm which presses against a piezeo-electric crystal (Figure 2.26). Movement of the diaphragm causes the crystal to be compressed and so charges produced on its surface. The crystal can be considered to be a capacitor which becomes charged as a result of the diaphragm movement and so a potential difference appears across it. If the pressure keeps the diaphragm at a particular displacement, the resulting electrical charge is not maintained but leaks away. Thus the sensor is not suitable for static pressure measurements. Typically such a sensor can be used for pressures up to about 1000 MPa with a non-linearity error of about ±1.9% of the full range value.

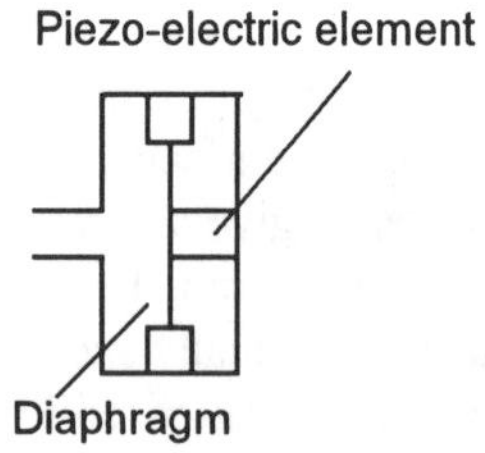

Piezo-electric element

Diaphragm

Figure 2.26 *Basic form of piezo-electric sensor*

2.2.4 Fluid flow

The traditional methods used for the measurement of fluid flow involve devices based on Bernoulli's equation. These involve the measurement of a pressure difference, the Venturi tube and the orifice plate described below being common examples. More modern methods have, however, been developed which more rapidly and efficiently record the flow rate and often with less interference to the flow.

Differential pressure methods

There are a number of methods commonly used for the measurement of the flow rate of fluids, or which the orifice plate is probably the most used,

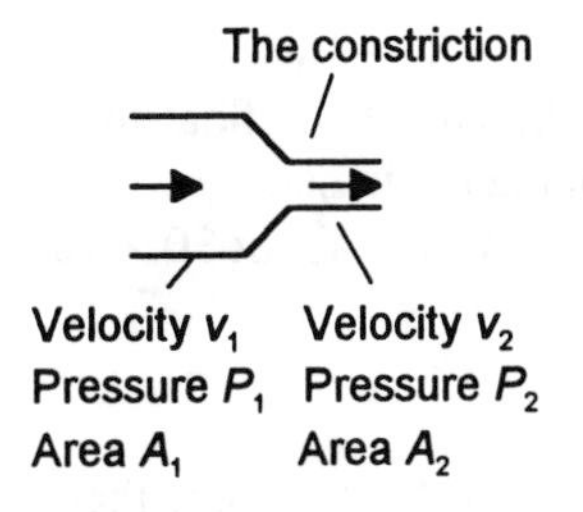

Figure 2.27 *Flow through a constriction*

based on the measurement of the pressure drop occurring when the fluid flows through a constriction (Figure 2.27). For a horizontal tube, where v_1 is the fluid velocity, P_1 the pressure and A_1 the cross-sectional area of the tube prior to the constriction, v_2 the velocity, P_1 the pressure and A_2 the cross-sectional area at the constriction, then Bernoulli's equation gives

$$\frac{v_1^2}{2g} + \frac{P_1}{\rho g} = \frac{v_2^2}{2g} + \frac{P_2}{\rho g}$$

with ρ as the liquid density and g the acceleration due to gravity. Since the mass of liquid passing per second through the tube prior to the constriction must equal that passing through the tube at the constriction, we have $A_1 v_1 \rho = A_2 v_2 \rho$. But the rate of flow, i.e. quantity Q of liquid passing through the tube per second, is $Q = A_1 v_1 = A_2 v_2$. Hence

$$Q = \frac{A_2}{\sqrt{1-(A_2/A_1)^2}} \sqrt{\frac{2(P_1 - P_2)}{\rho}}$$

The quantity of fluid flowing through the pipe per second is proportional to √(pressure difference). Hence, a measurement of the pressure difference can be used to give a measure of the rate of flow. There are many devices based on this principle, and the following example of the orifice plate is probably one of the most common.

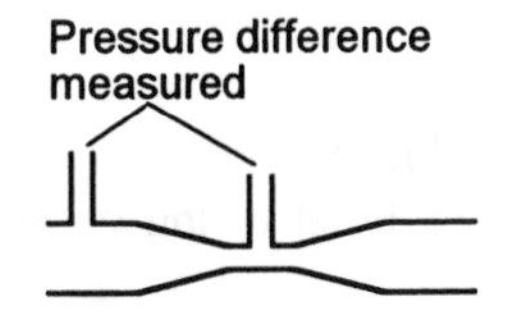

Figure 2.28 *The Venturi tube*

The *Venturi tube* is a tube which gradually tapers from the full pipe diameter to the constricted diameter. Figure 2.28 shows the typical form of such a tube. The pressure difference is measured between the flow prior to the construction and at the constriction, a diaphragm pressure cell generally being used. The device can be used for liquids containing particles, is capable of accuracy of better than ±0.5%, but does give a pressure loss of about 10 to 15% as a result of its presence in the flow line.

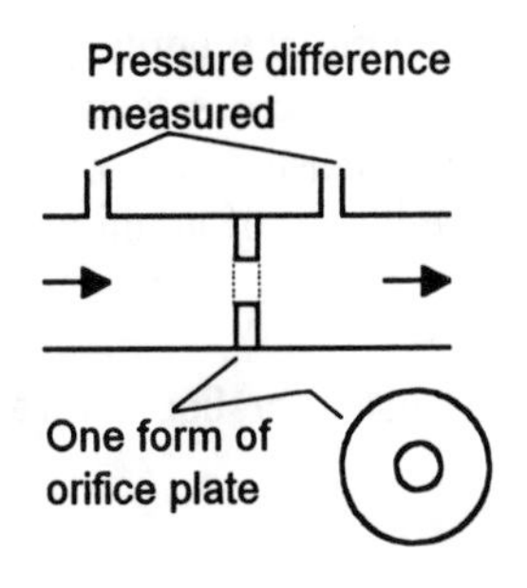

Figure 2.29 *Orifice plate*

The *orifice plate* (Figure 2.29) is simply a disc, with generally a central hole. The orifice plate is placed in the tube through which the fluid is flowing and the pressure difference measured between a point equal to the diameter of the tube upstream and a point equal to half the diameter downstream. Because of the way the fluid flows through the orifice plate, such measurements are equivalent to those taken with the Venturi tube. The orifice plate is simple, cheap, with no moving parts, has an accuracy of about ±1.5% of full range, and produces quite an appreciable pressure loss in the system to which it is connected. It does not work well with liquids containing particles because the orifice can become clogged.

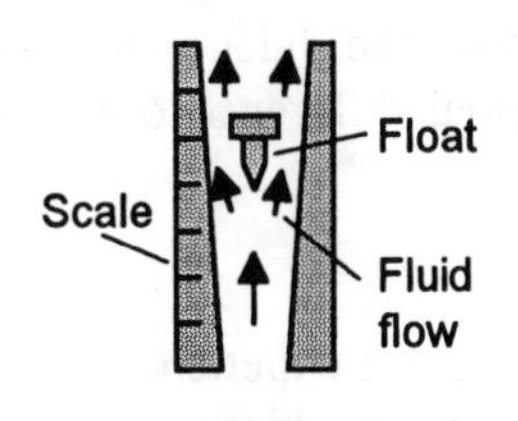

Figure 2.30 *Rotameter*

Rotameter

This consists of a float in a tapered vertical tube with the fluid flow pushing the float upwards (Figure 2.30). The fluid has to flow through the gap between the float and the walls of the tube. The tube is tapered and so the gap between the float and the tube walls increases as the float moves up the tube. This is an application of Bernouilli's equation since a pressure drop is produced by the fluid flowing through the gap and the float moves up to the

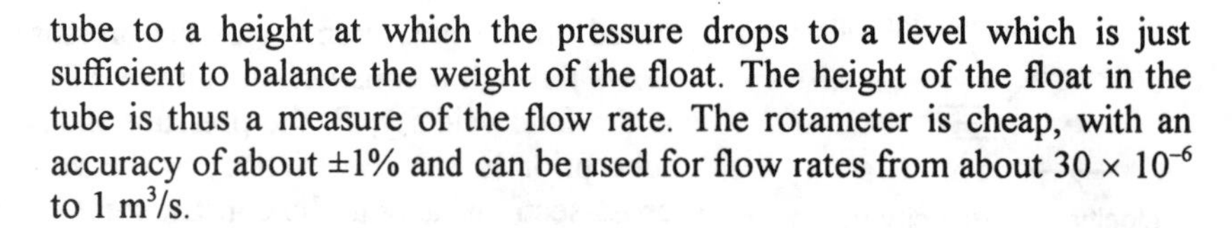

tube to a height at which the pressure drops to a level which is just sufficient to balance the weight of the float. The height of the float in the tube is thus a measure of the flow rate. The rotameter is cheap, with an accuracy of about ±1% and can be used for flow rates from about 30×10^{-6} to 1 m^3/s.

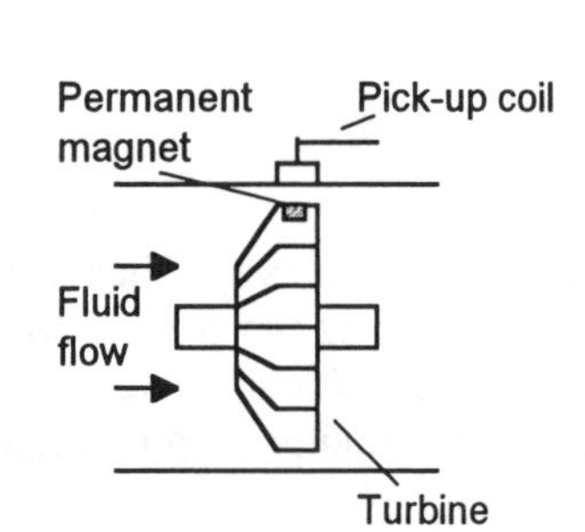

Figure 2.31 *Basic principle of the turbine flowmeter*

Turbine meter

The turbine flowmeter (Figure 2.31) consists of a multi-bladed rotor that is supported centrally in the pipe along which the flow occurs. The rotor rotates as a result of the fluid flow, the angular velocity being approximately proportional to the flow rate. The rate of revolution of the rotor can be determined by attaching a small permanent magnet to one of the blades and using a pick-up coil. An induced e.m.f. pulse is produced in the coil every time the magnet passes it. The pulses are counted and so the number of revolutions of the rotor can be determined. The meter is expensive, with an accuracy of typically about ±0.1%. Another form uses helical screws which rotate as a result of the fluid flow.

2.2.5 Liquid level

Methods used to measure the level of liquid in a vessel include those based on:

1 Floats whose position is directly related to the liquid level.
2 Archimedes' principle and a measurement of the upthrust acting on an object partially immersed in the liquid.
3 A measurement of the pressure at some point in the liquid, the pressure due to a column of liquid of height h being $h\rho g$, where ρ is the liquid density and g the acceleration due to gravity.
4 A measurement of the weight of the vessel containing the liquid plus liquid. The weight of the liquid is $Ah\rho g$, where A is the cross-sectional area of the vessel, h the height of liquid, ρ its density and g the acceleration due to gravity and thus changes in the height of liquid give weight changes.

The following give examples of the above methods used for liquid level measurements.

Floats

Figure 2.32 shows a simple float system. The float is at one end of a pivoted rod with the other end connected to the slider of a potentiometer. Changes in level cause the float to move and hence move the slider over the potentiometer resistance track and so give a potential difference output related to the liquid level.

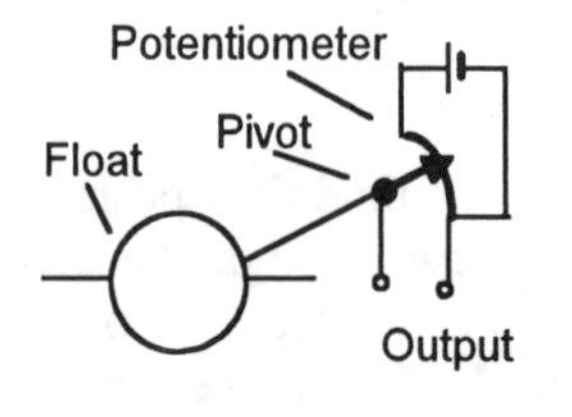

Figure 2.32 *Potentiometer float gauge*

Displacer gauge

When an object is partially or wholly immersed in a fluid it experiences an upthrust force equal to the weight of fluid displaced by the object. This is known as *Archimedes' principle*. Thus a change in the amount of an object

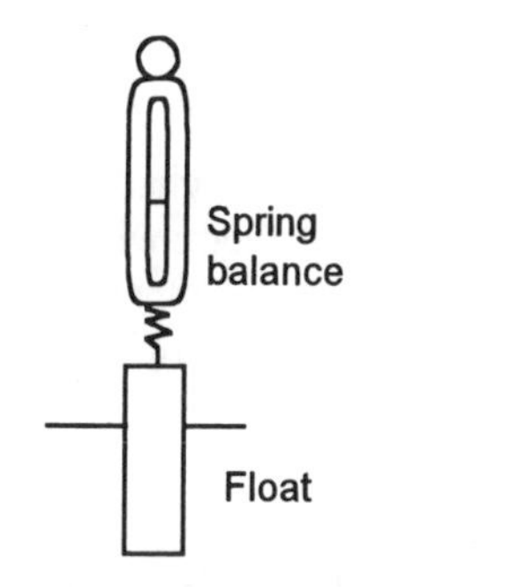

Figure 2.33 *Displacer gauge*

below the surface of a liquid will result in a change in the upthrust. The resultant force acting on such an object is then its weight minus the upthrust and thus depends on the depth to which the object is immersed. For a vertical cylinder of cross-sectional area A in a liquid of density ρ, if a height h of the cylinder is below the surface then the upthrust is $hA\rho g$, where g is the acceleration due to gravity, and so the apparent weight of the cylinder is $(mg - hA\pi g)$, where m is the mass of the cylinder. Displacer gauges need calibrating for liquid level determinations for particular liquids since the upthrust depends on the liquid density. Figure 2.33 shows a simple version of a displacement gauge.

Differential pressure

The pressure due to a height h of liquid above some level is $h\rho g$, where ρ is the liquid density and g the acceleration due to gravity. With a tank of liquid open to the atmosphere, the pressure difference can be measured between a point near the base of the tank and the atmosphere. The result is then proportional to the height of liquid above the pressure measurement point (Figure 2.34(a)). With a closed tank, the pressure difference has to be measured between a point near the bottom of the tank and in the gases above the liquid surface (Figure 2.34(b)). The pressure gauges used for such measurements tend to be diaphragm instruments.

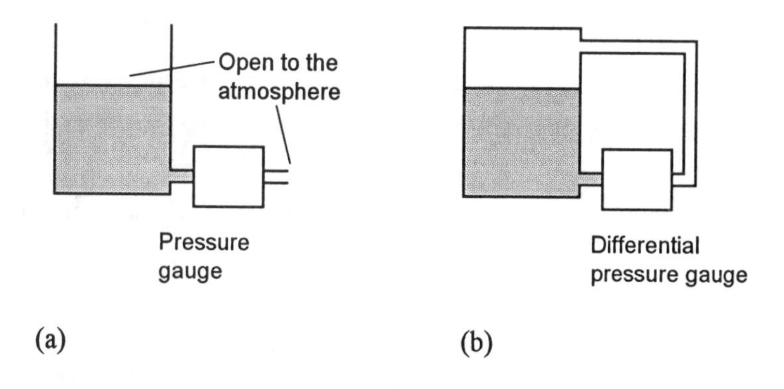

Figure 2.34 *Pressure level gauges*

Load cell

The weight of a tank of liquid can be used as a measure of the height of liquid in the tank. Load cells are commonly used for such weight measurements. Typically, a load cell consists of a strain gauged cylinder (Figure 2.35) which is included in the supports for the tank of liquid. When the level of the liquid changes, the weight changes and so the load on the load cell changes and the resistances of the strain gauges change. The resistance changes of the strain gauges are thus a measure of the level of the liquid. Since the load cells are completely isolated from the liquid, the method is useful for corrosive liquids.

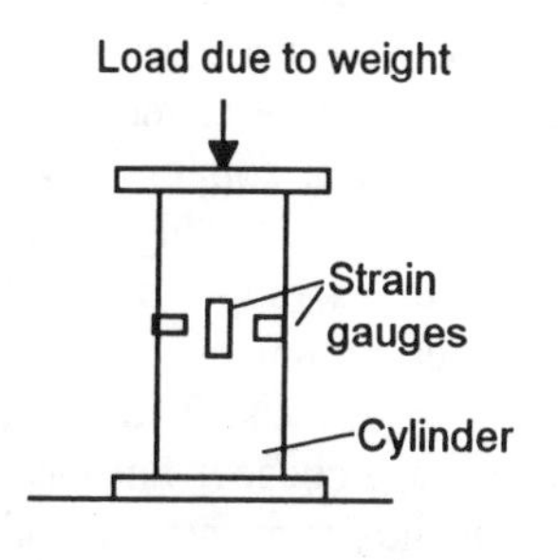

Figure 2.35 *Load cell*

2.2.6 Temperature

The expansion or contraction of solids, liquids or gases, the change in electrical resistance of conductors and semiconductors, and thermoelectric e.m.f.s are all examples of properties that change when the temperature changes and can be used as basis of temperature sensors. The following are some of the more commonly used temperature sensors.

Bimetallic strips

This device consists of two different metal strips of the same length bonded together (Figure 2.36). Because the metals have different coefficients of expansion, when the temperature increases the composite strip bends into a curved strip, with the higher coefficient metal on the outside of the curve. The amount by which the strip curves depends on the two metals used, the length of the composite strip and the change in temperature. If one end of a bimetallic strip is fixed, the amount by which the free end moves is a measure of the temperature. This movement may be used to open or close electric circuits, as in the simple thermostat commonly used with domestic heating systems. Bimetallic strip devices are robust, relatively cheap, have an accuracy of the order of ±1% and are fairly slow reacting to changes in temperature.

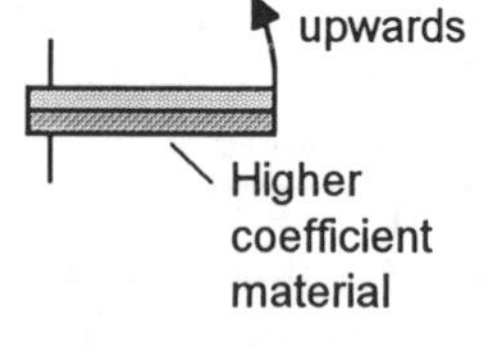

Figure 2.36 *Bimetallic strip*

Liquid in glass thermometers

The volume of a liquid increases when the temperature increases. The liquid in glass thermometer involves a liquid expanding up a capillary tube. The height to which the liquid expands is thus a measure of the temperature. With mercury as the liquid, the range possible is −35°C to +600°C, with alcohol −80°C to +70°C, with pentane −200°C to +30°C. Such thermometers are direct reading, fragile, capable of reasonable accuracy under standardised conditions, fairly slow reacting to temperature changes, and cheap.

Resistance temperature detectors (RTDs)

The resistance of most metals increases in a reasonably linear way with temperature (Figure 2.37) and can be represented by the equation:

$$R_t = R_0(1 + \alpha t)$$

where R_t is the resistance at a temperature t°C, R_0 the resistance at 0°C and α a constant for the metal, termed the temperature coefficient of resistance. Resistance temperature detectors (RTDs) are simple resistive elements in the form of coils of wire of such metals as platinum, nickel or copper alloys. Detectors using platinum have high linearity, good repeatability, high long term stability, can give an accuracy of ±0.5% or better, a range of about −200°C to +850°C, can be used in a wide range of environments without deterioration, but are more expensive than the other metals. They are, however, very widely used. Nickel and copper alloys are cheaper but have less stability, are more prone to interaction with the environment and cannot be used over such large temperature ranges.

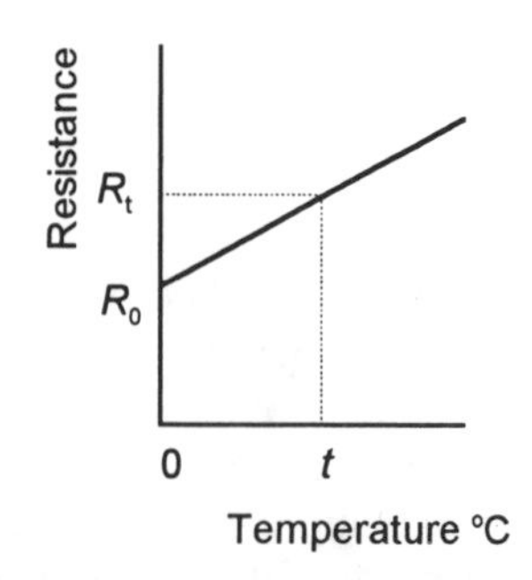

Figure 2.37 *Resistance variation with temperature for metals*

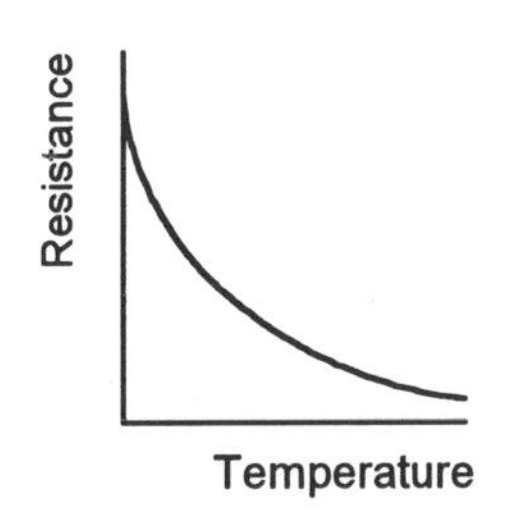

Figure 2.38 *Variation of resistance with temperature for thermistors*

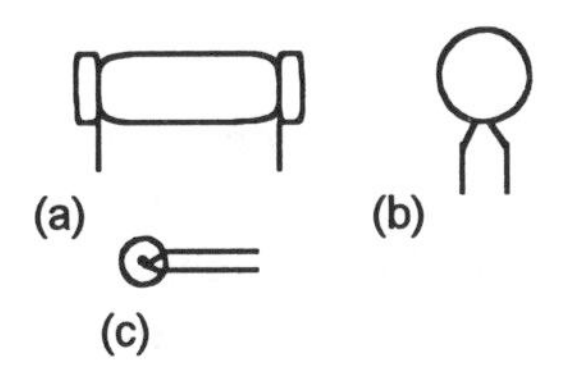

Figure 2.39 *(a) Rod, (b) disc, (c) bead thermistors*

Thermistors
Thermistors are semiconductor temperature sensors made from mixtures of metal oxides, such as those of chromium, cobalt, iron, manganese and nickel. The resistance of thermistors decreases in a very non-linear manner with an increase in temperature, Figure 2.38 illustrating this. The change in resistance per degree change in temperature is considerably larger than that which occurs with metals. For example, a thermistor might have a resistance of 29 kΩ at −20°C, 9.8 kΩ at 0°C, 3.75 kΩ at 20°C, 1.6 kΩ at 40°C, 0.75 kΩ at 60°C. The material is formed into various forms of element, such as beads, discs and rods (Figure 2.39). Thermistors are rugged and can be very small, so enabling temperatures to be monitored at virtually a point. Because of their small size they have small thermal capacity and so respond very rapidly to changes in temperature. The temperature range over which they can be used will depend on the thermistor concerned, ranges within about −100°C to +300°C being possible. They give very large changes in resistance per degree change in temperature and so are capable, over a small range, of being calibrated to give an accuracy of the order of 0.1°C or better. However, their characteristics tend to drift with time. Their main disadvantage is their non-linearity.

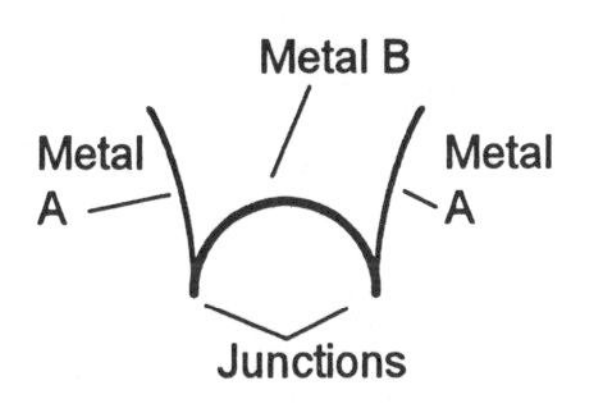

Figure 2.40 *Thermocouple*

Thermocouples
When two different metals are joined together, a potential difference occurs across the junction. The potential difference depends on the two metals used and the temperature of the junction. A thermocouple involves two such junctions, as illustrated in Figure 2.40. If both junctions are at the same temperature, the potential differences across the two junctions cancel each other out and there is no net e.m.f. If, however, there is a difference in temperature between the two junctions, there is an e.m.f. The value of this e.m.f. E depends on the two metals concerned and the temperatures t of both junctions. Usually one junction is held at 0°C and then, to a reasonable extent, the following relationship holds:

$$E = at + bt^2$$

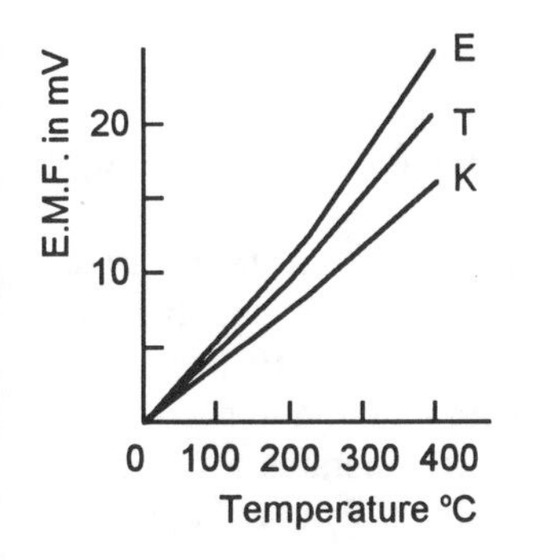

Figure 2.41 *Thermocouples, chromel-constantan (E), chromel-alumel (K), copper-constantan (T)*

where a and b are constants for the metals concerned. Figure 2.41 shows how the e.m.f. varies with temperature for a number of commonly used pairs of metals. Standard tables giving the e.m.f.s at different temperatures are available for the metals usually used for thermocouples. Commonly used thermocouples are listed in Table 2.1, with the temperature ranges over which they are generally used and typical sensitivities. These commonly used thermocouples are given reference letters. The base-metal thermocouples, E, J, K and T, are relatively cheap but deteriorate with age. They have accuracies which are typically about ±1 to 3%. Noble-metal thermocouples, e.g. R, are more expensive but are more stable with longer life. They have accuracies of the order of ±1% or better. Thermocouples are generally mounted in a sheath to give them mechanical and chemical protection. The response time of an unsheathed thermocouple is very fast. With a sheath this may be increased to as much as a few seconds if a large sheath is used.

Table 2.1 Thermocouples

Type	Materials	Range °C	Sensitivity µV/°C
E	Chromel–constantan	0 to 980	63
J	Iron–constantan	–180 to 760	53
K	Chromel–alumel	–180 to 1260	41
R	Platinum–platinum/ rhodium 13%	0 to 1750	8
T	Copper–constantan	–180 to 370	43

A thermocouple can be used with the reference junction at a temperature other than 0°C. However, the standard tables assume that the junction is at 0°C junction and hence a correction has to be applied before the tables can be used. The correction is applied using what is known as the *law of intermediate temperatures*, namely:

$$E_{t,0} = E_{t,I} + E_{I,0}$$

The e.m.f. $E_{t,0}$ at temperature t when the cold junction is at 0°C equals the e.m.f. $E_{t,I}$ at the intermediate temperature I plus the e.m.f. $E_{I,0}$ at temperature I when the cold junction is at 0°C. Consider a type E thermocouple. The following is data from standard tables.

Temp. (°C)	0	20	200
e.m.f. (mV)	0	1.192	13.419

Using the law of intermediate temperatures, the thermoelectric e.m.f. at 200° C with the cold junction at 20°C is:

$$E_{200,20} = E_{200,0} - E_{20,0} = 13.419 - 1.192 = 12.227 \text{ mV}$$

Note that this is not the e.m.f. given by the tables for a temperature of 180°C with a cold junction at 0°C, namely 11.949 mV.

To maintain one junction of a thermocouple at 0°C, it needs to be immersed in a mixture of ice and water. This, however, is often not convenient. A compensation circuit can, however, be used to provide an e.m.f. which varies with the temperature of the cold junction in such a way that when it is added to the thermocouple e.m.f. it generates a combined e.m.f. which is the same as would have been generated if the cold junction had been at 0°C (see section 2.4.3).

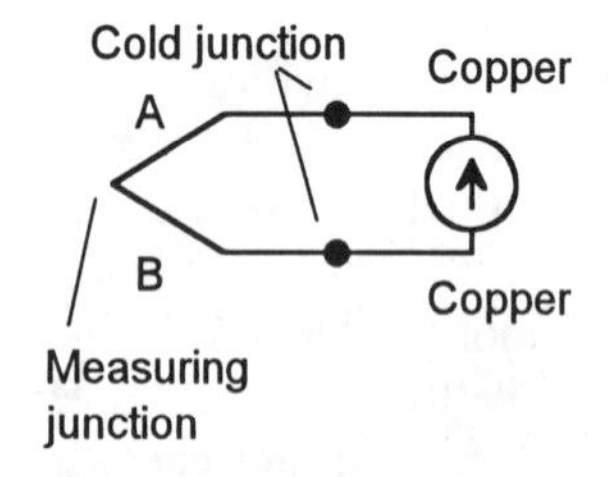

Figure 2.42 *The junctions with a measurement instrument*

When a thermocouple is connected to a measuring circuit, other metals can be involved (Figure 2.42). Thus we can have as the 'hot' junction that between metals A and B and the 'cold' junction effectively extended by the introduction of copper leads and the measurement instrument. Provided the junctions with the intermediate materials are at the same temperature, there is no extra e.m.f. involved and we still have the e.m.f. as due to the junction between metals A and B. The cold junction is thus the junctions between metal A and copper, and B and copper. These have both to be kept at 0°C if no compensation circuit is used.

2.3 Sensor selection

The selection of a sensor for a particular application requires a consideration of:

1 The nature of the measurement required, i.e. the sensor input. This means considering the variable to be measured, its nominal value, the range of values, the accuracy required, the required speed of measurement, the reliability required and the environmental conditions under which the measurement is to be made.

2 The nature of the output required from the sensor, this determining the signal processing required. The selection of sensors cannot be taken in isolation from a consideration of the form of output that is required from the system after signal processing, and thus there has to be a suitable marriage between sensor and signal processing.

Then possible sensors can be identified, taking into account such factors as their range, accuracy, linearity, speed of response, reliability, life, power supply requirements, ruggedness, availability and cost.

Example

Select a sensor which can be used to monitor the temperature of a liquid in the range 10°C to 80°C to an accuracy of about 1°C and which will give an output which can be used to change the current in an electrical circuit.

There are a number of forms of sensor that can be used to monitor such a temperature in the range, and to the accuracy, indicated. The choice is, however, limited by the requirement for an output which can change the current in an electrical circuit. This would suggest a resistance thermometer. In view of the limited accuracy and range required, a thermistor might thus be considered.

Example

Select a sensor which can be used for the measurement of the level of a corrosive acid in a circular vessel of diameter 1 m and will give an electrical output. The acid level can vary from 0 to 3 m and the minimum change in level to be detected is 0.1 m. The empty vessel has a weight of 50 kg. The acid has a density of 1050 kg/m^3.

Because of the corrosive nature of the acid there could be problems in using a sensor which is inserted in the liquid. Thus a possibility is to use a load cell, or load cells, to monitor the weight of the vessel. Such cells would give an electrical output. The weight of the liquid changes from 0 when empty to, when full, $1050 \times 3 \times \pi(1^2/4) \times 9.8 = 24.3$ kN. Adding this to the weight of the empty vessel gives a weight that varies from about 0.5 kN to 4.9 kN. A change of level of 0.1 m gives a change in weight of $0.10 \times 1050 \times \pi(1^2/4) \times 9.8 = 0.8$ kN. If the load of the vessel is spread between three load cells, each will require a range of about 0 to 5 kN with a resolution of about 0.3 kN.

2.4 Signal processing

The output signal from the sensor of a measurement system has generally to be processed in some way to make it suitable for display or use in some control system. For example, the signal may be too small and have to be amplified, be analogue and have to be made digital, be digital and have to be made analogue, be a resistance change and have to be made into a current change, be a voltage change and have to be made into a suitable size current change, be a pressure change and have to be made into a current change, etc. All these changes can be referred to as *signal processing*. For example, the output from a thermocouple is a very small voltage, a few millivolts. A signal processing module might then be used to convert this into a larger voltage and provide cold junction compensation (i.e. allow for the cold junction not being at 0°C).

The following sections outlines some of the elements that might be used in signal processing.

2.4.1 Resistance to voltage converter

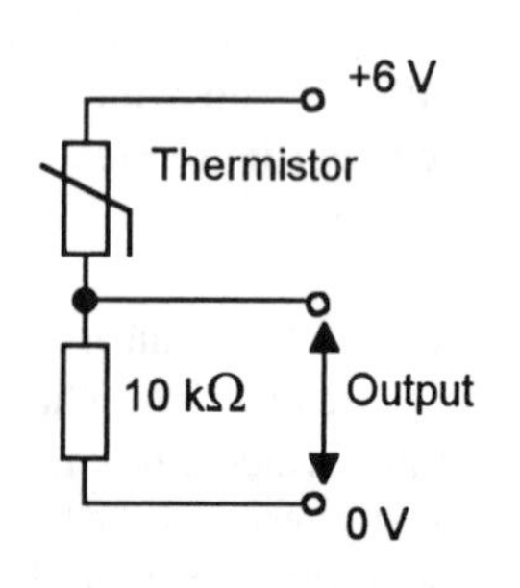

Figure 2.43 *Thermistor*

Consider how the resistance change produced by a thermistor when subject to a temperature change can be converted into a voltage change. Figure 2.43 shows how a *potential divider circuit* can be used. A constant voltage, of perhaps 6 V, is applied across the thermistor and another resistor in series. With a thermistor with a resistance of 4.7 kΩ, the series resistor might be 10 kΩ. The output signal is the voltage across the 10 kΩ resistor. When the resistance of the thermistor changes, the fraction of the 6 V across the 10 kΩ resistor changes.

The output voltage is proportional to the fraction of the total resistance which is between the output terminals. Thus:

$$\text{output} = \frac{R}{R + R_t} V$$

where V is the total voltage applied, in Figure 2.42 this is shown as 6 V, R the value of the resistance between the output terminals (10 kΩ) and R_t the resistance of the thermistor at the temperature concerned. The potential divider circuit is thus an example of a simple resistance to voltage converter. Another example of such a converter is the Wheatstone bridge.

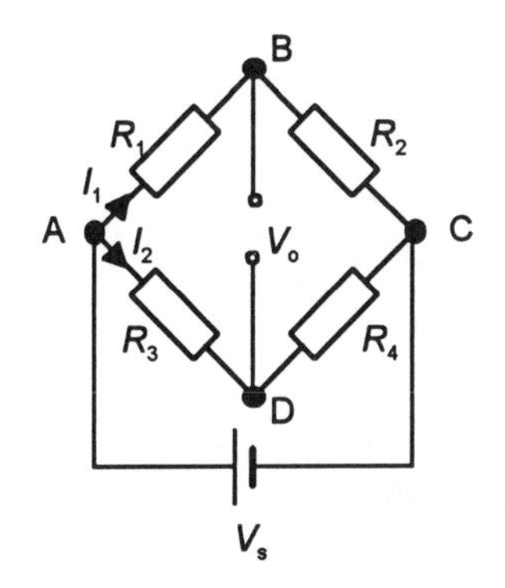

Figure 2.44 *The Wheatstone*

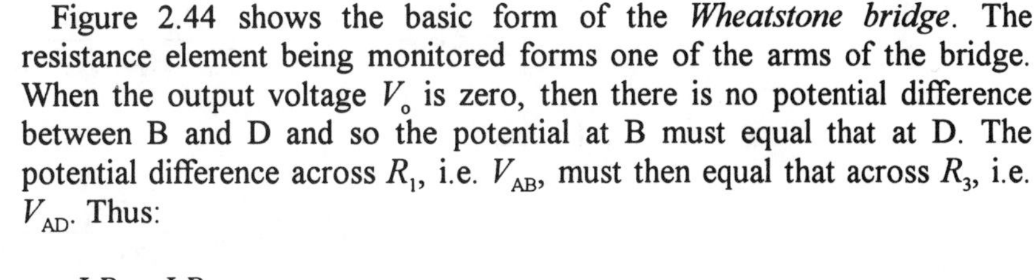

Figure 2.44 shows the basic form of the *Wheatstone bridge*. The resistance element being monitored forms one of the arms of the bridge. When the output voltage V_o is zero, then there is no potential difference between B and D and so the potential at B must equal that at D. The potential difference across R_1, i.e. V_{AB}, must then equal that across R_3, i.e. V_{AD}. Thus:

$$I_1R_1 = I_2R_2$$

We also must have the potential difference across R_2, i.e. V_{BC}, equal to that across R_4, i.e. V_{DC}. Since there is no current through BD then the current through R_2 must be the same as that through R_1 and the current through R_4 the same as that through R_3. Thus:

$$I_1R_2 = I_2R_4$$

Dividing these two equations gives:

$$\frac{R_1}{R_2} = \frac{R_3}{R_4}$$

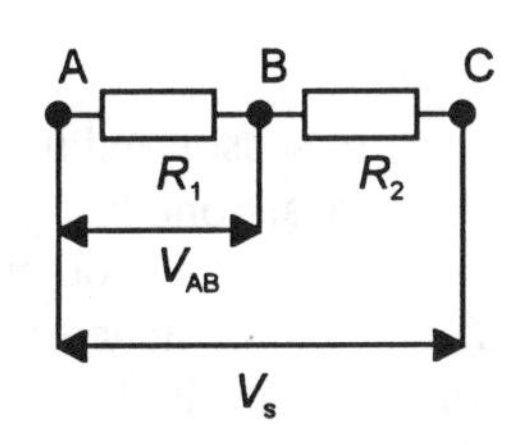

Figure 2.45 *Potential drop across* R_1

The bridge is said to be *balanced*.

Now consider what happens when one of the elements has a resistance which changes from this balanced condition. The supply voltage V_s is connected between points A and C and thus the potential drop across the resistor R_1 is the fraction $R_1/(R_1 + R_2)$ of the supply voltage (Figure 2.45). Hence:

$$V_{AB} = \frac{V_sR_1}{R_1 + R_2}$$

Similarly, the potential difference across R_3 (Figure 2.46) is:

$$V_{AD} = \frac{V_sR_3}{R_3 + R_4}$$

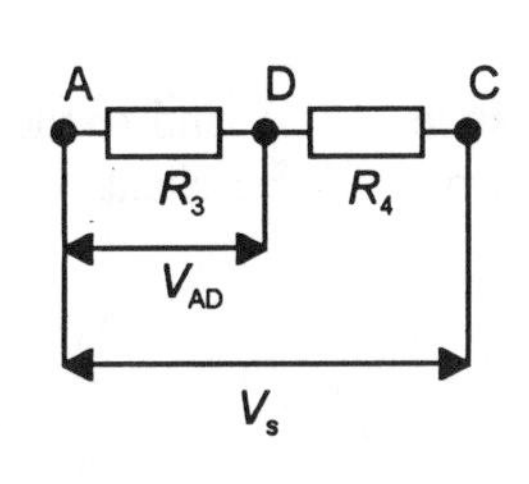

Figure 2.46 *Potential drop across* R_3

Thus the difference in potential between B and D, i.e. the output potential difference V_o, is:

$$V_o = V_{AB} - V_{AD} = V_s\left(\frac{R_1}{R_1 + R_2} - \frac{R_3}{R_3 + R_4}\right)$$

This equation gives the balanced condition when $V_o = 0$.

Consider resistance R_1 to be a sensor which has a resistance change, e.g. a strain gauge which has a resistance change when strained. A change in resistance from R_1 to $R_1 + \delta R_1$ gives a change in output from V_o to $V_o + \delta V_o$, where:

$$V_o + \delta V_o = V_s\left(\frac{R_1 + \delta R_1}{R_1 + \delta R_1 + R_2} - \frac{R_3}{R_3 + R_4}\right)$$

Hence:

$$(V_o + \delta V_o) - V_o = V_s\left(\frac{R_1 + \delta R_1}{R_1 + \delta R_1 + R_2} - \frac{R_1}{R_1 + R_2}\right)$$

If δR_1 is much smaller than R_1 then the denominator $R_1 + \delta R_1 + R_2$ approximates to $R_1 + R_2$ and so the above equation approximates to:

$$\delta V_o \approx V_s\left(\frac{\delta R_1}{R_1 + R_2}\right)$$

With this approximation, the change in output voltage is thus proportional to the change in the resistance of the sensor. We thus have a resistance to voltage converter. Note that the above equation only gives the output voltage when there is no load resistance across the output. If there is such a resistance then the loading effect has to be considered (see section 1.3).

Example

A platinum resistance coil is to be used as a temperature sensor and has a resistance at 0°C of 100 Ω. It forms one arm of a Wheatstone bridge with the bridge being balanced at this temperature and each of the other arms also being 100 Ω. If the temperature coefficient of resistance of platinum is 0.0039 K^{-1}, what will be the output voltage from the bridge per degree change in temperature if the supply voltage is 6.0 V?

The variation of the resistance of the platinum with temperature can be represented by:

$$R_t = R_0(1 + \alpha t)$$

where R_t is the resistance at t °C, R_0 the resistance at 0°C and α the temperature coefficient of resistance. Thus, for a one degree change in temperature:

$$\text{Change in resistance} = R_t - R_0 = R_0\alpha t$$

$$= 100 \times 0.0039 \times 1 = 0.39\ \Omega$$

Since this resistance change is small compared to the 100 Ω, the approximate equation developed above for the output voltage can be used. Hence the change in output per degree change in temperature is:

$$\delta V_o \approx V_s\left(\frac{\delta R_1}{R_1 + R_2}\right) = \frac{6.0 \times 0.39}{100 + 100} = 0.012\ \text{V}$$

2.4.2 Temperature compensation

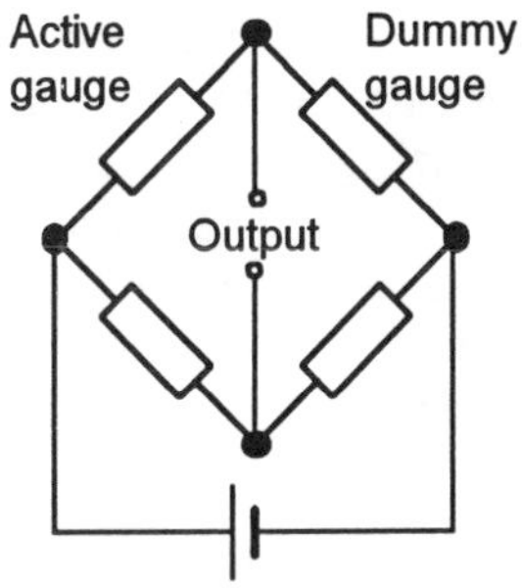

Figure 2.47 *Temperature compensation*

The electrical resistance strain gauge is a resistance element which changes resistance when subject to strain. However, it will also change resistance when subject to a temperature change. Thus, in order to use it to determine strain, compensation has to be made for temperature effects. One way of eliminating the temperature effect is to use a *dummy strain gauge*. This is a strain gauge identical to the one under strain, the active gauge, which is mounted on the same material as the active gauge but not subject to the strain. It is positioned close to the active gauge so that it suffers the same temperature changes. As a result, a temperature change will cause both gauges to change resistance by the same amount. The active gauge is mounted in one arm of a Wheatstone bridge (Figure 2.47) and the dummy gauge in an opposite arm so that the effects of temperature-induced resistance changes cancel out.

Strain gauges are often used with other sensors such as diaphragm pressure gauges or load cells. Temperature compensation is still required. While dummy gauges could be used, a better solution is to use four strain gauges with two of them attached so that the applied forces put them in tension and the other two in compression. The gauges, e.g. gauges 1 and 2, that are in tension will increase in resistance while those in compression, gauges 3 and 4, will decrease in resistance. The gauges are connected as the four arms of a Wheatstone bridge (Figure 2.48). As all the gauges and so all the arms of the bridge will be equally affected by any temperature changes the arrangement is temperature compensated. The arrangement also gives a much greater output voltage than would occur with just a single active gauge.

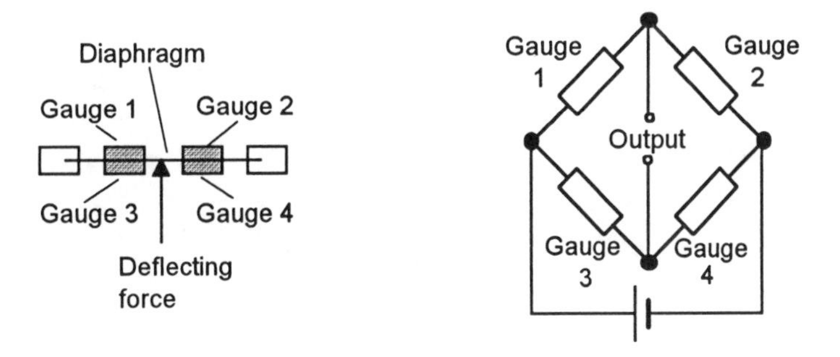

Figure 2.48 *Temperature compensation with four active strain gauges*

2.4.3 Thermocouple compensation

With a thermocouple, one junction should be kept at 0°C; the temperature can then be obtained by looking up in tables the e.m.f. produced by the thermocouple. Figure 2.49 illustrates what is required. This keeping of a junction at 0°C, i.e. in a mixture of ice and water, is not always feasible or very convenient and the cold junction is often allowed to be at the ambient

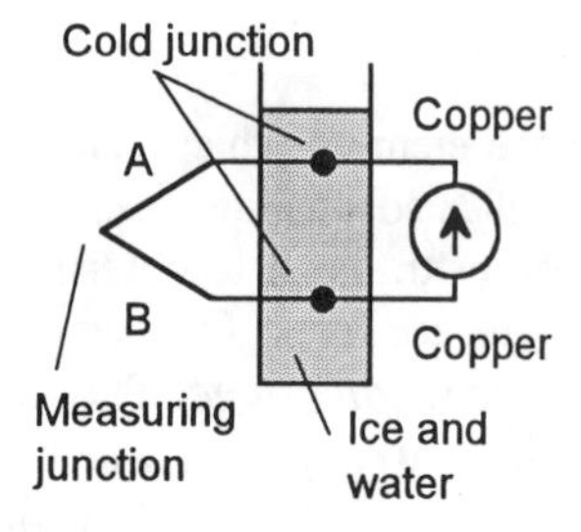

Figure 2.49 *Cold junction at 0°C*

temperature. To take account of this a compensation voltage has to be added to the thermocouple (see the law of intermediate temperatures in the item on thermocouples in section 2.2.6). This voltage is the same as the e.m.f. that would be generated by the thermocouple with one junction at 0°C and the other at the ambient temperature. We thus need a voltage which will depend on the ambient temperature. Such a voltage can be produced by using a resistance temperature sensor in a Wheatstone bridge. Figure 2.50 illustrates this. The bridge is balanced at 0°C and the output voltage from the bridge provides the correction potential difference at other temperatures. By a suitable choice of resistance temperature sensor, the appropriate voltage can be obtained.

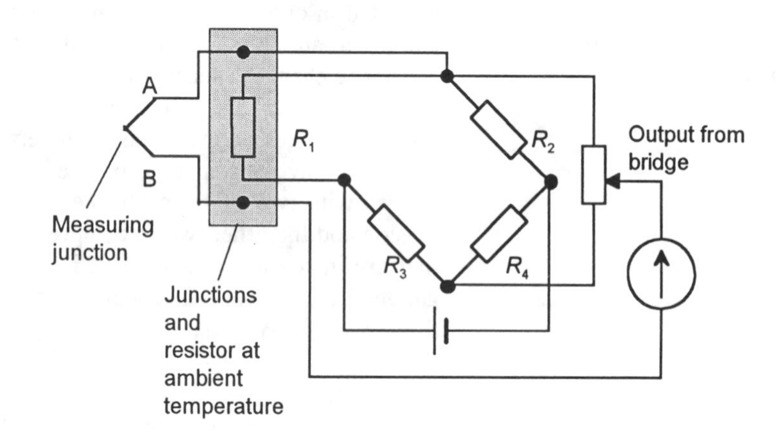

Figure 2.50 *A Wheatstone bridge compensation circuit*

The resistance of a metal resistance temperature sensor is given by:

$$R_t = R_0(1 + \alpha t)$$

where R_t is the resistance at t°C, R_0 the resistance at 0°C and α the temperature coefficient of resistance. Thus when there is a change in temperature,

$$\text{change in resistance} = R_t - R_0 = R_0\alpha t$$

The output voltage for the bridge, taking R_1 to be the resistance temperature sensor, is given by:

$$\delta V_o \approx V_s\left(\frac{\delta R_1}{R_1 + R_2}\right) = \frac{V_s R_0 \alpha t}{R_0 + R_2}$$

This voltage must be the same as that given by the thermocouple with one junction at 0°C and the other at the ambient temperature. The thermocouple e.m.f. e is likely to vary with temperature t in a reasonably linear manner over the small temperature range being considered, i.e. from 0°C to the ambient temperature. Thus we can write $e = at$, where a is a constant, i.e. the e.m.f. produced per degree change in temperature. Hence for compensation we must have:

$$at = \frac{V_s R_0 \alpha t}{R_0 + R_2}$$

and so the condition:

$$aR_2 = R_0(V_s\alpha - a)$$

Example

Determine the value of the resistance R_2 if compensation is to be provided for an iron–constantan thermocouple giving 51 μV/°C. The compensation is to be provided by a nickel resistance element with a resistance of 10 Ω at 0°C and a temperature coefficient of resistance of 0.0067 K^{-1}. Take the supply voltage for the bridge to be 2.0 V.

Using the equation developed above:

$$aR_2 = R_0(V_s\alpha - a)$$

then:

$$51 \times 10^{-6} \times R_2 = 10(2 \times 0.0067 - 51 \times 10^{-6})$$

Hence R_2 is 2617 Ω.

2.4.4 Protection

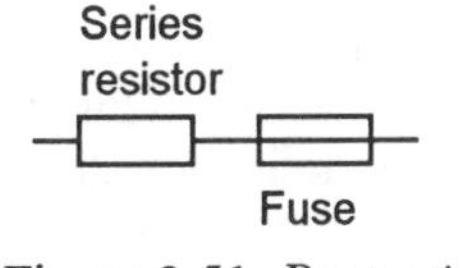

Figure 2.51 *Protection against high currents*

An important element that is often present with signal processing is protection against high currents or high voltages. For example, sensors when connected to a microprocessor can damage it if high currents or high voltages are transmitted from the sensor to the microprocessor. A high current can be protected against by the incorporation in the input line of a series resistor to limit the current to an acceptable level and a fuse to break if the current does exceed a safe level (Figure 2.51). Protection against high voltages and wrong polarity voltages may be obtained by the use of a Zener diode circuit (Figure 2.52). The Zener diode with a reverse voltage connected across it has a high resistance up to some particular voltage at which is suddenly breaks down and becomes conducting (Figure 2.52(a)). Zener diodes are given voltage ratings, the rating indicating at which voltage they become conducting. For example, to allow a maximum voltage of 5 V but stop voltages above 5.1 V being applied to the following circuit, a Zener diode with a voltage rating of 5.1 V might be chosen. For voltages below

5.1 V the Zener diode, in reverse voltage connection, has a high resistance. When the voltage rises to 5.1 V the Zener diode breaks down and its resistance drops to a very low value. Thus, with the circuit shown in Figure 2.52(b), with the applied voltage below 5.1 V, the Zener diode, in reverse voltage connection, has a much higher resistance than the other resistor and so virtually all the applied voltage is across the Zener diode. When the applied voltage rises to 5.1 V, the Zener diode breaks down and has a low resistance. As a consequence, most of the voltage is then dropped across the resistor and the voltage across the diode drops. Thus the output voltage drops. Because the Zener diode is a diode with a low resistance for current in one direction through it and a high resistance for the opposite direction, it also provides protection against wrong polarity.

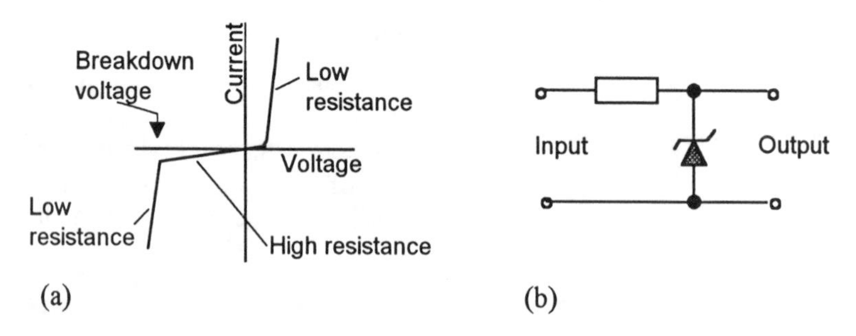

Figure 2.52 *(a) Current-voltage relationship for a Zener diode, (b) protection circuit*

To ensure protection, it is often necessary to completely isolate circuits so that there are no electrical connections between them. This can be done using an *optoisolator*. Such a device converts an electrical signal into an optical signal, transmits it to a detector which then converts it back into an electrical signal (Figure 2.53). The input signal passes through an infrared light-emitting diode (LED) and so produces a beam of infrared radiation. This infrared signal is then detected by a phototransistor. To prevent the LED having the wrong polarity or too high an applied voltage, it is likely to be protected by the Zener diode circuit (of the type shown above in Figure 2.52). Also, if there is likely to be an alternating signal in the input a diode would be put in the input line to rectify it.

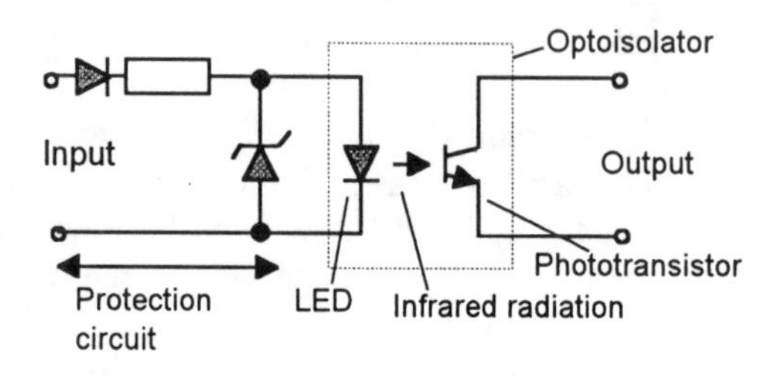

Figure 2.53 *An optoisolator*

2.4.5 Analogue-to-digital conversion

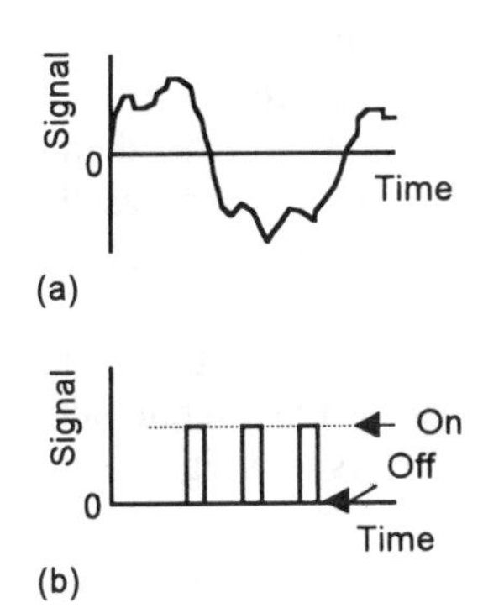

Figure 2.54 *(a) An analogue signal, (b) a digital signal*

The electrical output from sensors such as thermocouples, resistance elements used for temperature measurement, strain gauges, diaphragm pressure gauges, LVDTs, etc. is in analogue form. An analogue signal (Figure 2.54(a)) is one that is continuously variable, changing smoothly over a range of values. The signal is an analogue, i.e. a scaled version, of the quantity it represents. A digital signal increases in jumps, being a sequence of pulses, often just on-off signals (Figure 2.54(b)). Digital circuits are switching type circuits which react to changes in signals from one value to another.

Microprocessors require digital inputs. Thus, where a microprocessor is used as part of a measurement or control system, the analogue output from a sensor has to be converted into a digital form before it can be used as an input to the microprocessor. The output from a microprocessor is digital. Most control elements require an analogue input and so the digital output from a microprocessor has to be converted into an analogue form before it can be used by them. Thus there is a need for analogue-to-digital converters (ADC) and digital-to-analogue converters (DAC). This section is a consideration of analogue-to-digital converters, the next section being digital-to-analogue converters.

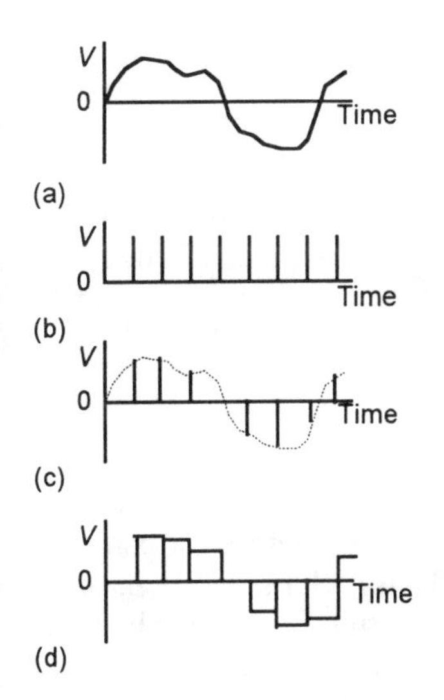

Figure 2.55 *(a) Analogue signal, (b) time signal, (c) sampled signal, (d) sampled and held signal*

Analogue-to-digital conversion involves a number of stages. The first stage is to take samples of the analogue signal (Figure 2.55(a)). A clock supplies regular time signal pulses (Figure 2.55(b)) to the analogue-to-digital converter and every time it receives a pulse it samples the analogue signal. The result is a series of narrow pulses with heights which vary in accord with the variation of the analogue signal (Figure 2.55(c)). This sequence of pulses is changed into the signal form shown in Figure 2.55(d) by each sampled value being held until the next pulse occurs. It is necessary to hold a sample of the analogue signal so that conversion can take place to a digital signal at an analogue-to-digital converter. This converts each sample into a sequence of pulses representing the value. For example, the first sampled value might be represented by 101, the next sample by 011, etc. The 1 represents an 'on' or 'high' signal, the 0 an 'off' or 'low' signal. Analogue-to-digital conversion thus involves a sample and hold unit followed by an analogue-to-digital converter (Figure 2.56).

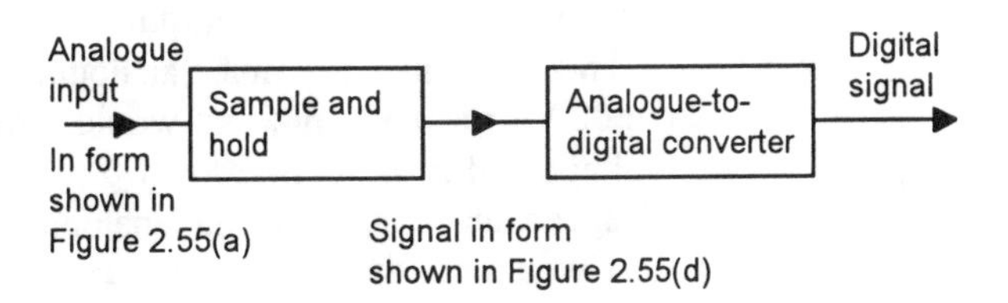

Figure 2.56 *Analogue-to-digital conversion*

To illustrate the action of the analogue-to-digital converter, consider one that gives an output restricted to three bits. The binary digits of 0 and 1, i.e. the 'low' and 'high' signals, are referred to as *bits*. A group of bits is called a *word*. Thus the three bits give the *word length* for this particular analogue-to-digital converter. The word is what represents the digital version of the analogue voltage. The position of bits in a word has the significance that the least significant bit is on the right end of the word and the most significant bit on the left. This is just like counting in tens, 435 has the 5 as the least significant number and the 4 as the most significant number, the least significant number contributing least to the overall value of the 435 number. The sequence of bits in a word of n bits thus signifies:

$2^{n-1}, \ldots, 2^3, 2^2, 2^1, 2^0$

Most significant bit — Least significant bit

With binary numbers we have the basic rules that:

$$0 + 0 = 0$$

$$0 + 1 = 1$$

$$1 + 1 = 10$$

Thus if we start with 000 and add 1 we obtain 001. If we add a further 1 we have 010. Adding another 1 gives 011. With three bits in a word we thus have the possible words of:

000 001 010 011 100 101 110 111

There are eight possible words which can be used to represent the analogue input. Thus we divide the maximum analogue voltage into eight parts and one of the digital words corresponds to each. Thus each rise in the analogue voltage of (1/8) of the maximum analogue input results in a further bit being generated. Thus for word 000 we have 0 V input. To generate the next digital word of 001 the input has to rise to 1/8 of the maximum voltage. To generate the next word of 010 the input has to rise to 2/8 of the maximum voltage. Figure 2.57 illustrates this conversion of the sampled and held input voltage to a digital output.

Thus if we had a sampled analogue input of 8 V, the digital output would be 000 for a 0 V input and would remain at that output until the analogue voltage had risen to 1 V, i.e. 1/8 of the maximum analogue input. It would then remain at 001 until the analogue input had risen to 2 V. This value of 001 would continue until the analogue input had risen to 3 V. The smallest change in the analogue voltage that would result in a change in the digital output is thus 1 V. This is termed the *resolution* of the converter.

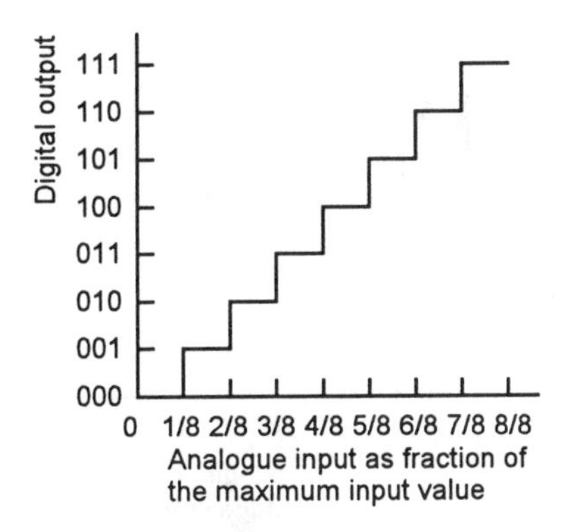

Figure 2.57 *Digital output from an analogue-to-digital converter*

The word length possible with an analogue-to-digital converter determines its *resolution*. With a word length of n bits the maximum, or full scale, analogue input V_{FS} is divided into 2^n pieces. The minimum change in input that can be detected, i.e. the *resolution*, is thus $V_{FS}/2^n$. With an analogue-to-digital converter having a word length of 10 bits and the maximum analogue signal input range 10 V, then the maximum analogue voltage is divided into $2^{10} = 1024$ pieces and the resolution is $10/1024 = 9.8$ mV.

There are a number of forms of analogue-to-digital converter. A basic form involves the analogue input voltage being compared with another analogue voltage that is steadily increased in steps. Figure 2.58(a) illustrates this when the voltage being converted is 6 V. When the increasing voltage reaches the same value at the input voltage the process is halted and a binary count made of the number of steps taken during the process. The comparison voltage is generated by a clock supplying pulses which are converted by a digital-to-analogue converter into a steadily increasing analogue voltage (Figure 2.58(b)). The binary count of the pulses then gives the digital conversion of the analogue input voltage.

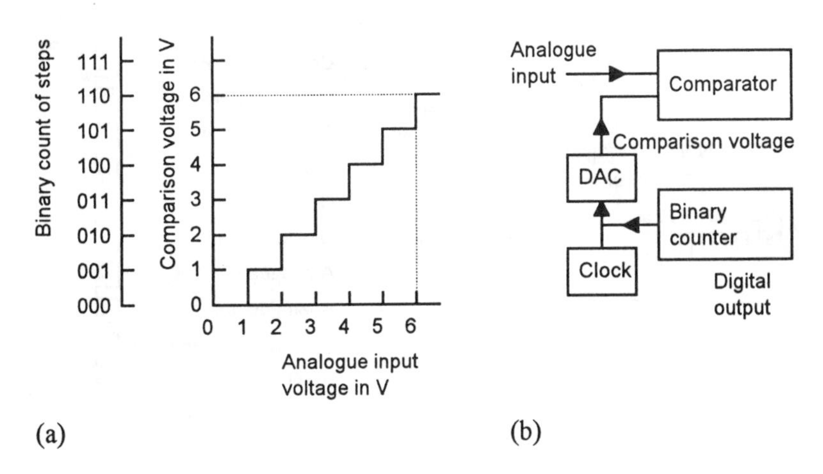

Figure 2.58 *The counter form of analogue-to-digital converter*

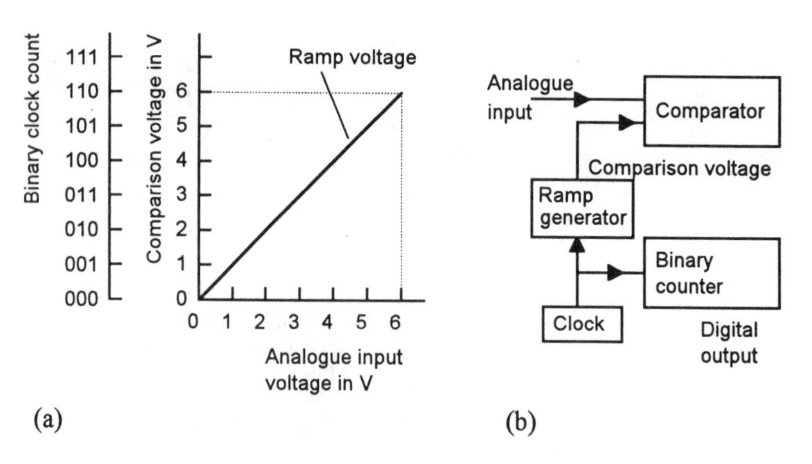

Figure 2.59 *The ramp form of analogue-to-digital converter*

Another form is the *ramp* form of analogue-to-digital converter. This involves an analogue voltage being increased at a constant rate, a so-called ramp voltage, and applied to a comparator where it is compared with the analogue voltage from the sensor. The time taken for the ramp voltage to increase to the value of the sensor voltage will depend on the size of the sampled analogue voltage. When the ramp voltage starts, a binary counter starts counting the regular pulses from a clock. When the two voltages are equal, the process stops and the binary word indicated by the counter is the digital representation of the sampled analogue voltage. Figure 2.59 indicates the subsystems involved in the ramp form of analogue-to-digital converter.

Analogue-to-digital converters are generally purchased as integrated circuits. Figure 2.60 shows an example of the pin connections for one. The term *conversion time* is used to specify the time it takes a converter to generate a complete digital word when supplied with the analogue input, that of the circuit in Figure 2.60 being 5 microseconds.

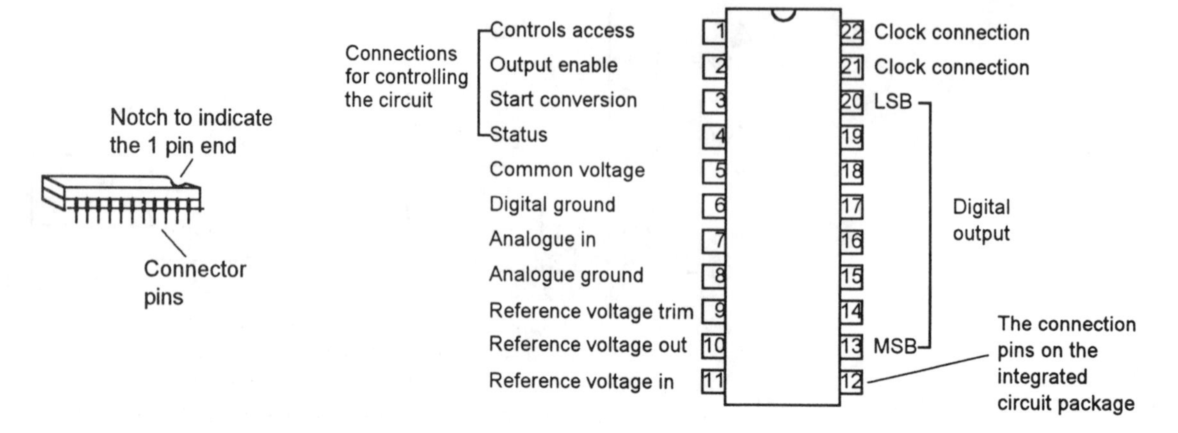

Figure 2.60 *The GEC Plessey ZN439E 8-bit analogue-to-digital converter*

Example

A thermocouple gives an output of 0.4 mV for each degree change in temperature. What will be the word length required when its output passes through an analogue-to-digital converter if temperatures from 0 to 200°C are to be measured with a resolution of 0.5°C?

The full scale output from the sensor is 200 × 0.4 = 80 mV. With a word length n there are 2^n digital numbers. Thus this voltage will be divided into 2^n levels and so the minimum voltage change that can be detected is $80/2^n$ mV. For a resolution of 0.5°C we must be able to detect a signal from the sensor of 0.5 × 0.4 = 0.20 mV. Hence:

$$0.20 = \frac{80}{2^n}$$

and so $2^n = 400$ and $n = 8.6$. Thus a 9-bit word length is required.

2.4.6 Digital-to-analogue converters

The input to a digital-to-analogue converter is a binary word and the output its equivalent analogue value. For example, if we have a full scale output of 7 V then a digital input of 000 will give 0 V, 001 give 1 V, ... and 111 the full scale value of 7 V. Figure 2.61 illustrates this.

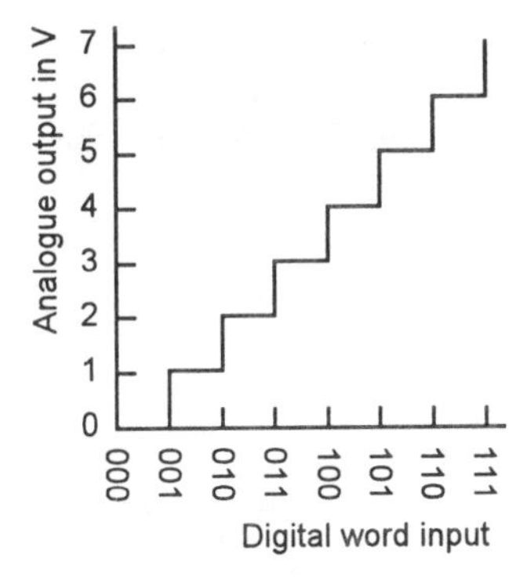

Figure 2.61 *Digital-to-analogue conversion*

The basic form of a digital-to-analogue converter involves the digital input being used to activate electronic switches such that a 1 activates a switch and a 0 does not, the position of the 1 in the word determining which switch is activated. Thus when, with say a 3-bit converter, 001 is received we have a voltage of, say, 1 V switched to the output, when 010 is received we have 2 V, switched to the output, and when 100 is received we have 4 V switched to the output. Hence if we have the digital word 011 we have the least significant bit 001 switching 1 V to the output and the 010 bit 2 V to the output to give a summed output of 3 V (Figure 2.62).

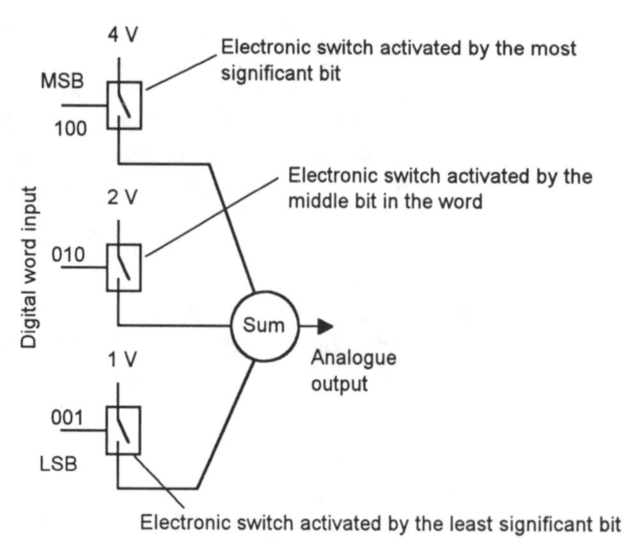

Figure 2.62 *The principle of a 3-bit digital-to-analogue converter*

One way this can be achieved is by using a *R-2R ladder network*. Figure 2.63 shows such a network for a three-bit digital-to-analogue converter. The output voltage is generated by switching sections of the ladder to either the reference voltage or 0 V according to whether there is a 1 or 0 in the digital input. The arrangement of series and parallel resistors ensures that the correct voltage is applied to each of the electronic switches. Digital-to-analogue converters can be purchased as integrated circuits. Figure 2.64 shows details of the GEC Plessey ZN558D 8-bit latched input digital-to-analogue converter. This contains the electronic switches and a *R-2R* ladder network. After the conversion is complete, the 8-bit result is placed in an internal latch until the next conversion is complete. A latch is just a device to retain the output until a new one replaces it.

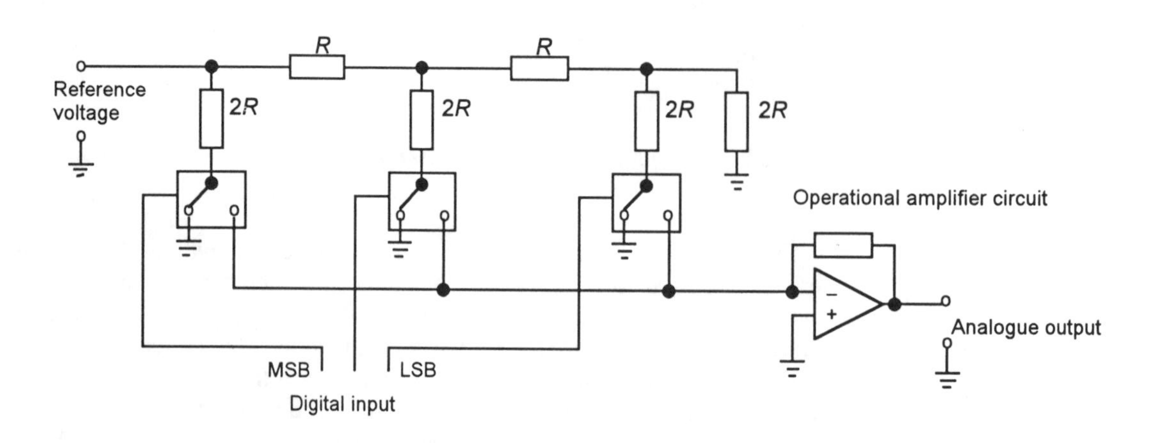

Figure 2.63 *An R-2R ladder network digital-to-analogue converter*

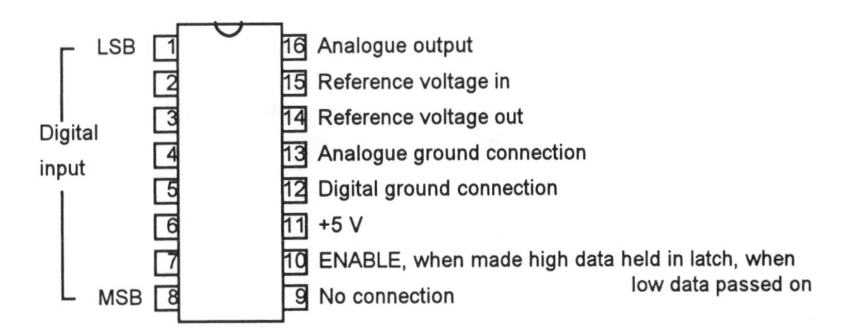

Figure 2.64 *The GEC Plessey ZN558D 8-bit latched digital-to-analogue converter*

Example

A microprocessor gives an output of an 8-bit word. This is fed through an 8-bit digital-to-analogue converter to a control valve which requires 6.0 V to be fully open. If the fully open state is to be indicated by the output of the digital word 11111111 what will be the change in output to the valve when there is a change of 1 bit?

The output voltage will be divided into 2^8 intervals. Since there is to be an output of 6.0 V when the output is 2^8 of these intervals, a change of 1 bit is a change in the output voltage of $6.0/2^8 = 0.023$ V.

2.4.7 Multiplexers

The analogue-to-digital converters described in section 2.4.5 are single-input devices, in that a converter just transforms a single analogue voltage into a digital signal. Frequently, however, there is a need for measurements to be sampled from a number of different locations, or perhaps a number of different measurements need to be made. Thus several analogue signals need to be converted to digital signals. While a separate analogue-to-digital converter could be used for each analogue signal, a cheaper alternative is to use an *analogue multiplexer*. Figure 2.65 shows how a multiplexer is used.

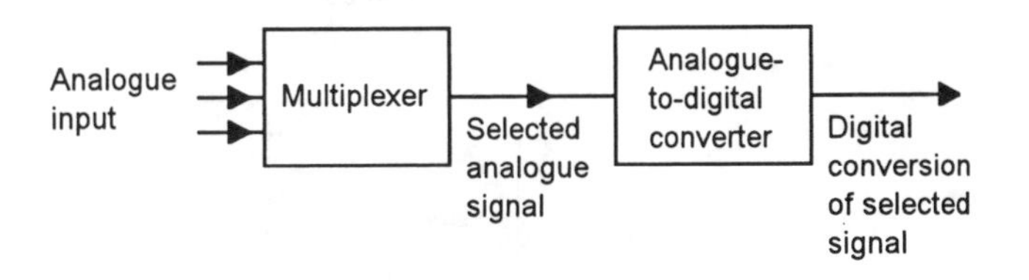

Figure 2.65 *Multiplexer use*

The multiplexer is essentially a switching device which enables each of the analogue inputs to be sampled in turn and connected to a single analogue-to-digital converter. Figure 2.66 shows the basic form of a multiplexer for three analogue inputs. Such circuits are available in the form of integrated circuits.

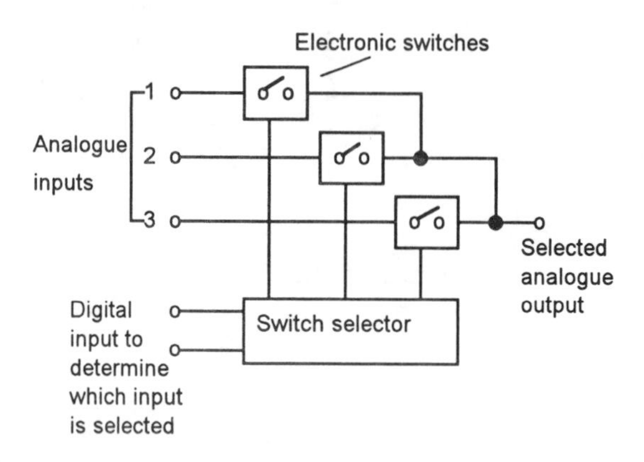

Figure 2.66 *Multiplexer principle*

2.4.8 Pressure-to-current converter

Control systems generally are electrical and so there is often a need to convert a pressure into an electrical current. Figure 2.67 shows the basic principle of such a converter. The input pressure causes the bellows to extend and so apply a force to displace the end of the pivoted beam. The movement of the beam results in the core being moved in a linear variable differential transformer (LVDT). This gives an electrical output which is amplified. The resulting current is then passed through a solenoid. The current through the solenoid produces a magnetic field which is used to attract the end of the pivoted beam to bring the beam back to its initial horizontal position. When the beam is back in the position, the solenoid current maintaining it in this position is taken as the converted form of the pressure input.

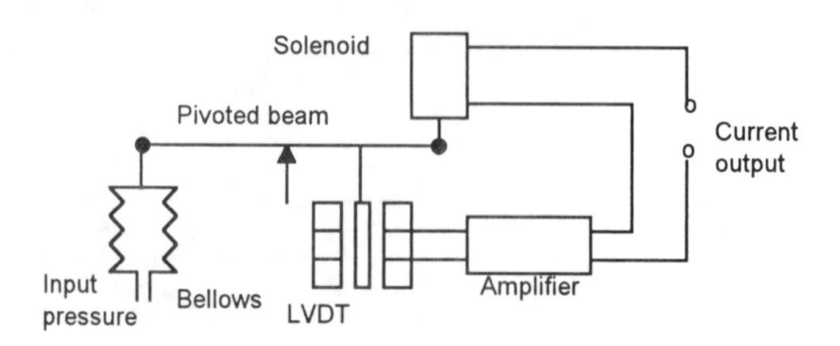

Figure 2.67 *Pressure-to-current converter*

2.5 Data presentation elements

The elements that can be used for the presentation of data can be classified into two groups: indicators and recorders. *Indicators* give an instant visual indication of the sensed variable while *recorders* record the output signal over a period of time and give automatically a permanent record. A recorder will be the most appropriate choice if the event is high speed or transient and cannot be followed by an observer, or there are large amounts of data, or it is essential to have a record of the data. Both indicators and recorders can be subdivided into two groups of devices, *analogue* and *digital*. An example of an analogue indicator is a meter which has a pointer moving across a scale, while a digital meter would be just a display of a series of numbers. An example of an analogue recorder is a chart recorder which has a pen moving across a moving sheet of paper, while a digital recorder has the output printed out on a sheet of paper as a sequence of numbers. Figure 2.68 illustrates the types of data presentation elements that can be classified in the above groups, commonly used methods being indicated in each group.

The following are some brief notes about some of the characteristics of commonly used data presentation methods.

Pointer moving over a fixed scale

The *moving coil meter* is an analogue data presentation element involving a pointer moving across a fixed scale. The basic instrument movement is a d.c. microammeter with shunts, multipliers and rectifiers being used to convert it to other ranges of direct current and alternating current, direct voltage and alternating voltage. With alternating current and voltages, the instrument is restricted to frequencies between about 50 Hz and 10 kHz. The overall accuracy is generally of the order of ±0.1 to ±5 %. The time taken for a moving coil meter to reach a steady deflection is typically in the region of a few seconds. The low resistance of the meter can present loading problems.

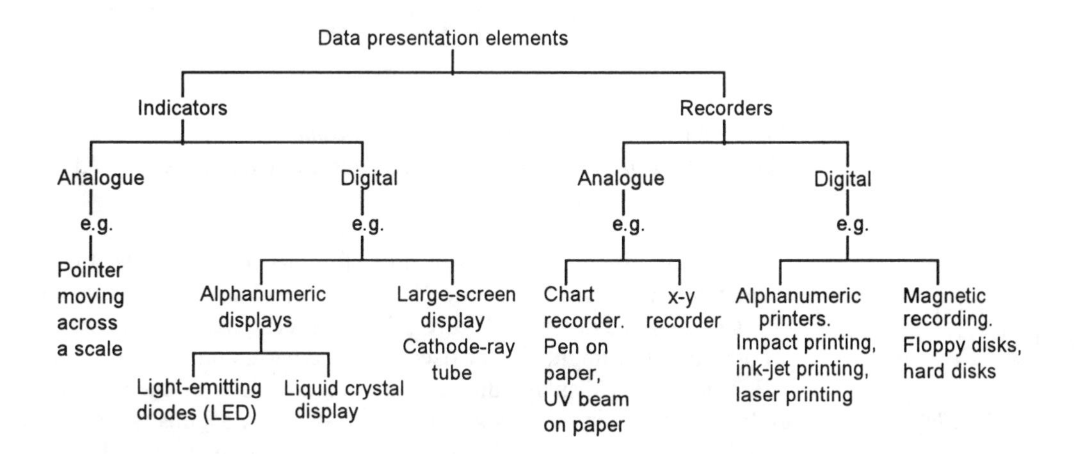

Figure 2.68 *Data presentation elements*

LED and liquid crystal displays

The term *alphanumeric display* is used for one that can display the letters of the alphabet and numbers. Such display systems generally use light-emitting diodes (LEDs) or liquid crystal displays. Light-emitting diodes require low voltages and low currents in order to emit light and are cheap. The most commonly used LEDs can give red, yellow or green colours. Two basic types of array are used to generate alphanumeric displays, segmented and dot matrix. The 7-segment display (Figure 2.69(a)) is a common form. By illuminating different segments of the display the full range of numbers and a small range of alphabetical characters can be formed. For example, to form a 2 the segments a, b, d, e and g are illuminated. The 5 × 7 dot matrix (Figure 2.69(b)) display enables a full range of numbers and alphabetical characters to be produced by illuminating different segments in a rectangular array.

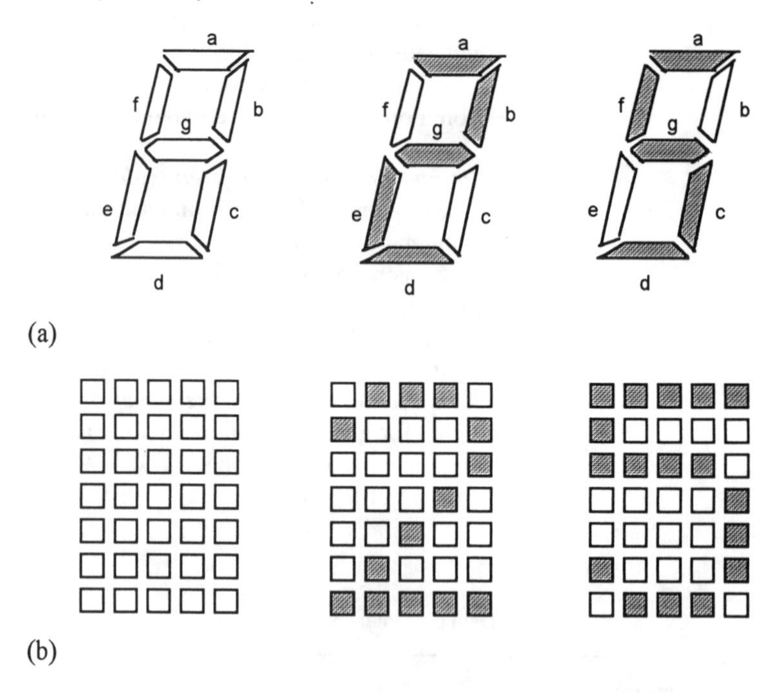

Figure 2.69 *(a) 7-segment format with examples of the numbers 2 and 5, (b) 7 × 5 dot-matrix format with examples of the numbers 2 and 5*

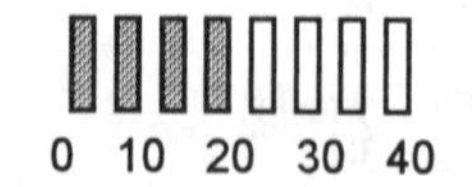

Figure 2.70 *Bar type of display*

LEDs can also be arranged in other formats. For example, they can be arranged in the form of bars, the length of the illuminated bar then being a measure of some quantity (Figure 2.70). A speedometer might use this form of display. They might also be used to give a bar graph form of display.

Liquid crystal displays do not produce any light of their own but use reflected light. Liquid crystals can be arranged in segments like the LEDs shown above. The crystal segments are on a reflecting plate. When an electric field is applied to a crystal, light is no longer passed through it and so there is no reflected light. That segment then appears dark.

An example of an instrument using such forms of display is the *digital voltmeter*. This voltmeter gives its reading in the form of a sequence of digits. The digital voltmeter is essentially just a sample and hold unit feeding an analogue-to-digital converter with its digital output counted and the count displayed (Figure 2.71). It has a high resistance, of the order of 10 MΩ, and so loading effects are less likely than with the moving coil meter with its lower resistance. The sample and hold unit takes samples and thus the specification of the sample rate with such an instrument gives the time taken for the instrument to process the signal and give a reading. Thus, if the input voltage is changing at a rate which results in significant changes during the sampling time the voltmeter reading can be in error.

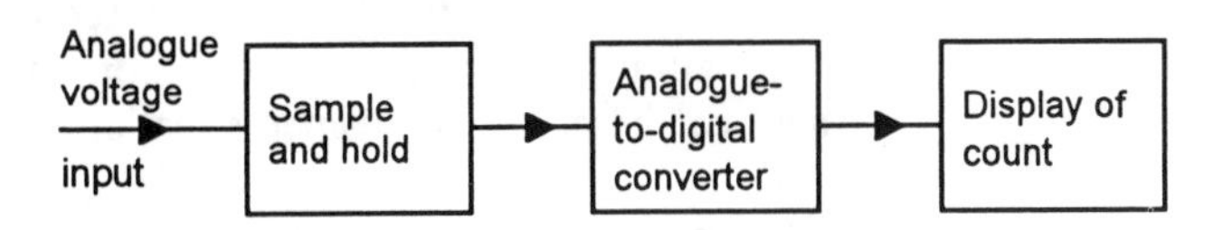

Figure 2.71 *Principle of the digital voltmeter*

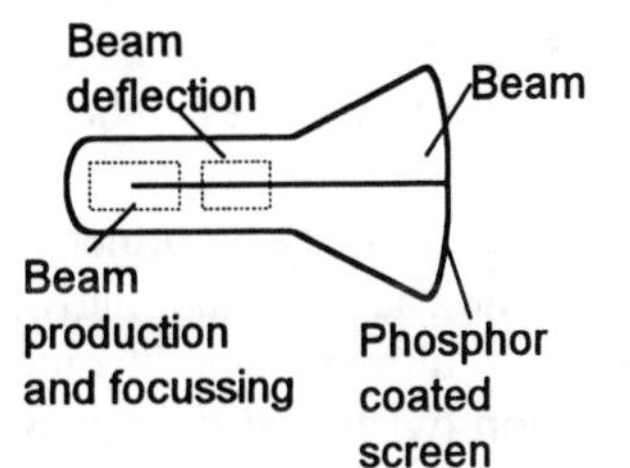

Figure 2.72 *Cathode ray tube*

Large screen display

Large screen displays can be obtained by the use of a cathode ray tube (Figure 2.72). Such a tube employs an electron beam which is focused on the inside surface of a screen which is coated with a large number of phosphor dots. The phosphor emits visible radiation in response to the impact of the electrons. To produce images on the screen, the electron beam is swept from left to right in a zig-zag pattern down the screen (Figure 2.73), the dots on the screen being made to light up according to whether the electron beam is switched on or off.

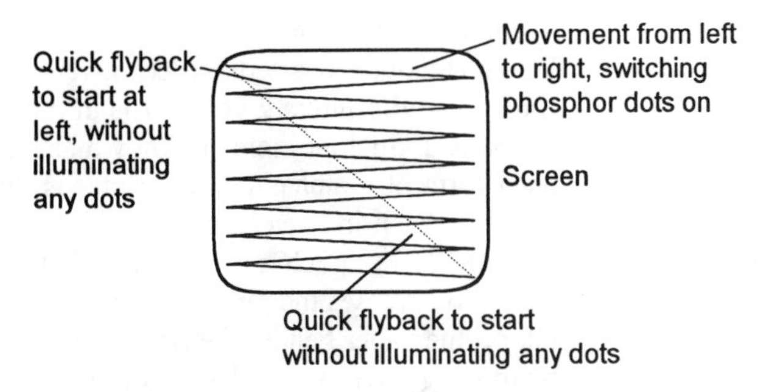

Figure 2.73 *Producing the screen display*

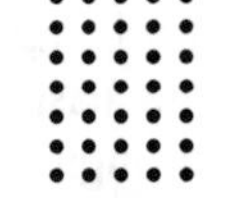

Figure 2.74 *7 by 5 dot matrix display*

The screen display consists of a large number of small picture elements, i.e. the phosphor dots, which are switched on or off. These elements are called *pixels*. Each character might, for example, be represented by a 7 by 5 matrix of dots (Figure 2.74). Figure 2.75 shows how such characters might be built up by the movement of the electron beam from left to right across

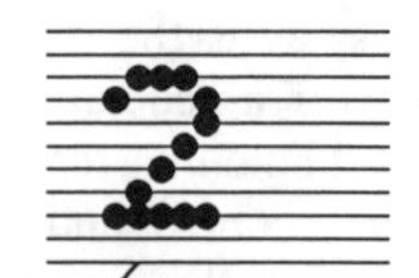

Figure 2.75 *Build up of characters*

the screen. The display might be a fixed format one with alphanumeric characters or a graphic display in which graphical images, i.e. pictures, being displayed in fixed positions, e.g. a diagram of part of a chemical plant with data such as measured values of pressure, etc. being displayed at appropriate points.

The *cathode-ray oscilloscope* is a voltage-measuring instrument using a cathode ray tube and which is capable of displaying extremely high frequency signals. The deflection of the electron beam is a measure of the input voltage. A general-purpose instrument can respond to signals up to about 10 MHz while more specialist instruments can respond up to about 1 GHz. Double-beam oscilloscopes enable two separate traces to be observed simultaneously on the screen while storage oscilloscopes enable the trace to remain on the screen after the input signal has ceased, only being removed by a deliberate action of erasure. Permanent records of traces can be made with cameras that attach directly to the oscilloscope.

Galvanometric chart recorder

The *galvanometric type* of chart recorder (Figure 2.76) works on the same principle as the moving coil meter movement. A coil is suspended between two fixed points by a suspension wire and in the magnetic field produced by a permanent magnet. When a current passes through the coil a torque acts on it, causing it to rotate and twist the suspension. The coil rotates to an angle at which the torque is balanced by the opposing torque resulting from the twisting of the suspension. The rotation of the coil results in a pen being moved across a chart.

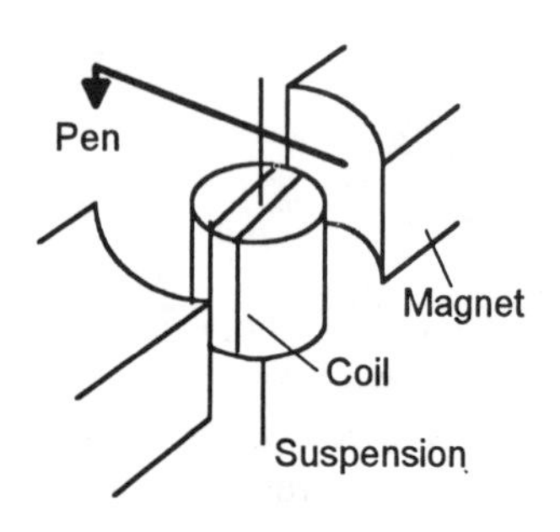

Figure 2.76 *The principle of the galvanometric chart*

The ultraviolet galvanometric chart recorder works on the same principle but instead of using a pointer moving a pen across the chart, a small mirror is attached to the suspension and reflects a beam of ultraviolet light onto sensitive paper. When the coil rotates, the suspension twists and the mirror rotates and so moves the beam across the chart.

Alphanumeric printer

While analogue chart recorders give records in the form of a continuous trace, digital printers give records in the form of numbers, letters or special characters. Such printers are known as *alphanumeric printers*. A commonly used form of a alphanumeric printer is the *dot-matrix printer*. With such a printer, the print head consists of either 9 or 24 pins in a vertical line (Figure 2.77). Each pin is controlled by an electromagnet which when turned on presses the pin onto the inking ribbon and so gives a small blob of ink on the paper behind the ribbon. The alphanumeric characters are formed by moving the print head across the paper and firing the appropriate pins.

Figure 2.77 *A 9-pin print head*

Magnetic recording

Digital data can be stored on magnetic tape or disks, the disks being referred to as *hard disks* or *floppy disks*. Figure 2.78 shows the basic principle of magnetic recording. The recording current is passed through a coil wrapped round a ferromagnetic core. This core has a small non-magnetic gap. The proximity of the magnetic tape or disk to the gap means that the magnetic flux in the core is readily diverted through it. The magnetic tape of disk consists of a plastic base coated with a ferromagnetic powder. When magnetic flux passes through this material it becomes permanently magnetised. Thus a magnetic record can be produced of the current through the coil.

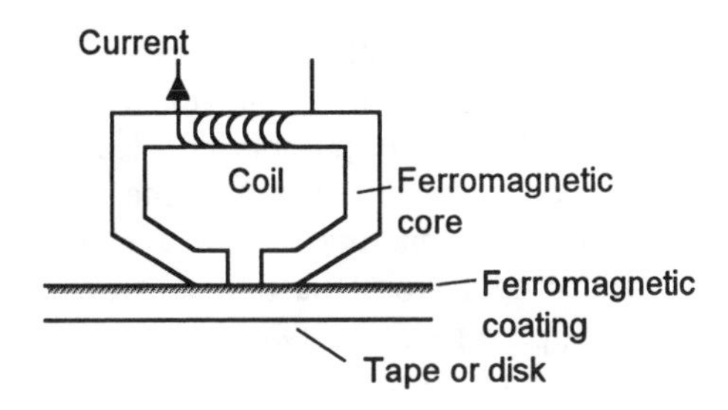

Figure2.78 *A magnetic recording (write) head*

The patterns of magnetism on a tape or disk can be read by passing it under a similar head to the recording head. The movement of the magnetised material under the head results in magnetic flux passing through the core of the head and electromagnetic induction producing a current through the coil wrapped round the head.

Hard disks and floppy disks operate in much the same way. The data is stored on the disk surface along concentric circles called tracks, a single disk having many such tracks. A single read/write head is used for each disk surface and the heads are mechanically moved to access the different tracks. The disk is spun by the hard disk or floppy disk drive and the read/write heads read or write data into a track. Figure 2.79 shows the basic forms of hard disks and floppy disks. The 3½ inch floppy disk used in the personal computer has 135 tracks per inch and can store 1.4 Mbytes of data.

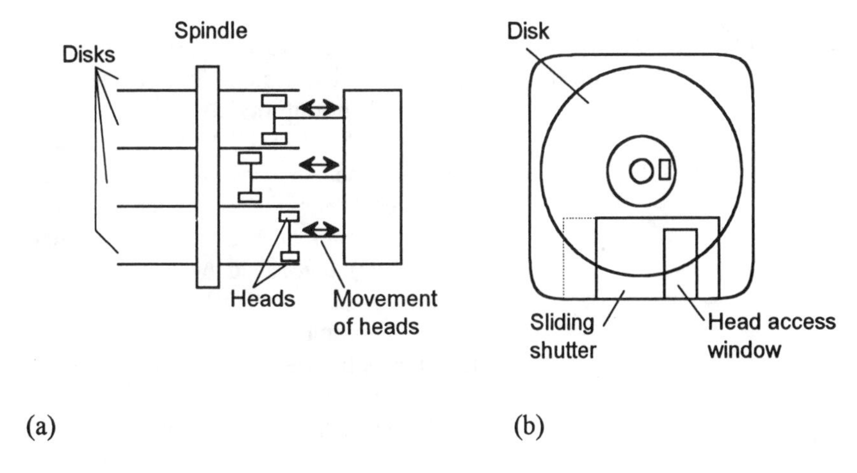

Figure 2.79 *Computer disks: (a) hard, (b) floppy*

2.5.1 Data loggers

The term *data logger* is used for a data-acquisition system for which the functions can be programmed from a front panel on the instrument. A data logger for handling analogue inputs from sensors consists essentially of a multiplexer, a sample and hold element and an analogue-to-digital converter (Figure 2.80). Such a unit can monitor the inputs from a large number of sensors. Inputs from individual sensors, after suitable signal conditioning, are fed into the multiplexer which is then used to select a signal from one sensor. This is then fed to the sample and hold element which holds the signal long enough for the analogue-to-digital conversion to occur. Thus the output from the unit is a digital signal for one of the sensors. The multiplexer can be switched to each sensor in turn and so the output can consist of a sequence of samples from each sensor. Scanning of the inputs can be selected to be just a sampling of a single sensor, a single scan of all sensors, a continuous scan of all sensors, and periodic scan of all sensors, e.g. every 1, 5, 15, 30 or 60 minutes. The output from the system might be displayed on a digital meter that indicates the output and which sensor is providing the input, i.e. the input channel number, or be used to give a permanent record with a digital printer or a magnetic disk memory.

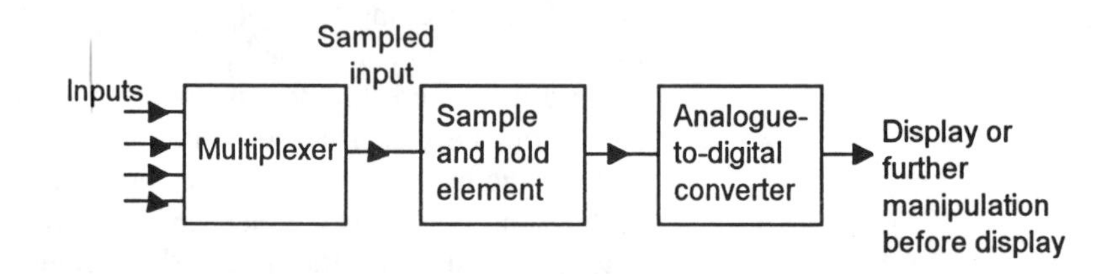

Figure 2.80 *A data logger*

2.6 Computer-based data acquisition

Figure 2.81 shows the basic elements of a typical microcomputer-based data acquisition system. The signals from the sensor are first processed to convert their outputs to a common signal range, typically to a voltage ranging from 0 to 5 V. A multiplexer is then used to select the signal from a particular sensor, the selection being controlled by signals from the computer. Thus the computer can be programmed to select signals at particular times or in a particular sequence. What we have is a data logger (see above) controlled by a computer. Since computers can only handle digital inputs, all the analogue signals have then to be converted into digital form. The computer might be programmed to react to, or further process, the signals before display. Thus, for example, it might be used to take the average of a number of readings and display just one value, or perhaps give an alarm signal if a value rises to above some specified limit.

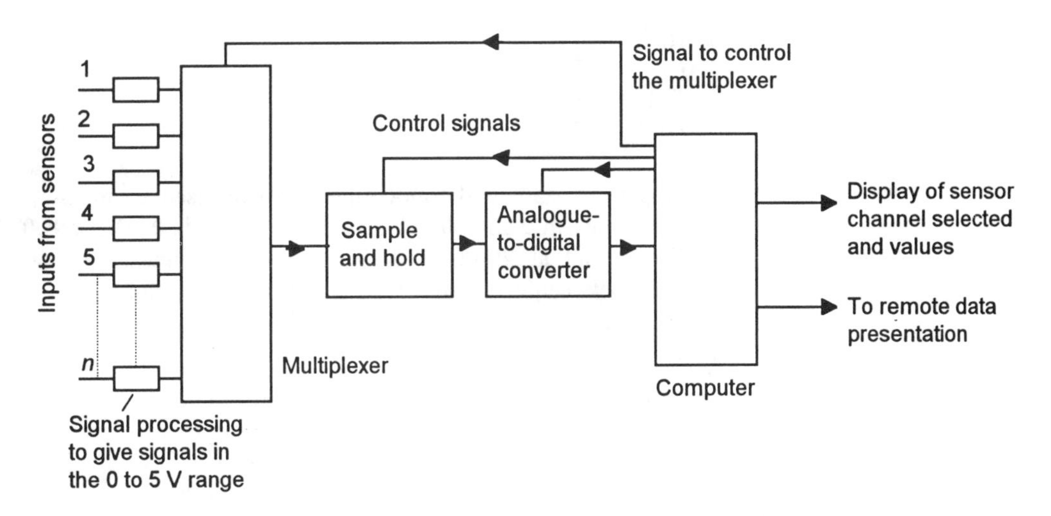

Figure 2.81 *Computer-based data acquisition*

Problems

Questions 1 to 20 have four answer options: A, B, C and D. Choose the correct answer from the answer options.

Questions 1 to 6 relate to the following information. The outputs from sensors can take a variety of forms. These include changes in:

A Displacement
B Resistance
C Voltage
D Capacitance

Select the output from above which is concerned with the following sensors:

1 A thermocouple which has an input of a temperature change.
2 A thermistor which has an input of a temperature change.
3 A diaphragm pressure cell which has an input of a change in the pressure difference between its two sides.
4 A LVDT which has an input of a change in displacement.
5 A strain gauge which has an input of a change in length.
6 A Bourdon gauge which has an input of a pressure change.

7 Decide whether each of these statements is True (T) or False (F).

In selecting a temperature sensor for monitoring a rapidly changing temperature, it is vital that the sensor has:
(i) A small thermal capacity.
(ii) High linearity.

Which option BEST describes the two statements?

A (i) T (ii) T
B (i) T (ii) F
C (i) F (ii) T
D (i) F (ii) F

8 A copper–constantan thermocouple is to be used to measure temperatures between 0 and 200°C. The e.m.f. at 0°C is 0 mV, at 100°C it is 4.277 mV and at 200°C it is 9.286 mV. If a linear relationship is assumed between e.m.f. and temperature over the full range, the non-linearity error at 100°C is:

A −3.9°C
B −7.9°C
C +3.9°C
D +7.9°C

9 The change in resistance of an electrical resistance strain gauge with a gauge factor of 2.0 and resistance 50 Ω when subject to a strain of 0.001 is:

A 0.0001 Ω
B 0.001 Ω
C 0.01 Ω
D 0.1 Ω

10 An incremental shaft encoder gives an output which is a direct measure of:

A The absolute angular position of a shaft.
B The change in angular rotation of a shaft.
C The diameter of the shaft.
D The change in dimater of the shaft.

11 A pressure sensor consisting of a diaphragm with strain gauges bonded to its surface has the following information in its specification:

Range: 0 to 1000 kPa
Non-linearity error: ±0.15% of full range
Hysteresis error: ±0.05% of full range

The total error due to non-linearity and hysteresis for a reading of 200 kPa is:

A ±0.2 kPa
B ±0.4 kPa
C ±2 kPa
D ±4 kPa

12 The water level in an open vessel is to be monitored by a diaphragm pressure cell responding to the difference in pressure between that at the base of the vessel and the atmosphere. The range of pressure differences across the diaphragm that the cell will have to respond to if the water level can vary between zero height above the cell measurement point and 1 m above it is (take the acceleration due to gravity to be 9.8 m/s^2 and the density of the water as 1000 kg/m^3):

A 102 Pa
B 102 kPa
C 9800 Pa
D 9800 kPa

13 Decide whether each of these statements is True (T) or False (F).

A float sensor for the determination of the level of water in a container is cylindrical with a mass 1.0 kg, cross-sectional area 20 cm^2 and a length of 0.5 m. It floats vertically in the water and presses upwards against a beam attached to its upward end.
(i) The maximum force that can act on the beam is 9.8 N.
(ii) The minimum force that can act on the beam is 8.8 N.

Which option BEST describes the two statements?

A (i) T (ii) T
B (i) T (ii) F
C (i) F (ii) T
D (i) F (ii) F

14 A Wheatstone bridge when used as a signal processing element can have an input of a change in resistance and an output of:

A A bigger resistance change.
B A digital signal.
C A voltage.
D A current.

15 The resolution of an analogue-to-digital converter with a word length of 8 bits and an analogue signal input range of 10 V is:

A 39 mV
B 625 mV
C 1.25 V
D 5 V

16 A sensor gives a maximum analogue output of 5 V. The word length is required for an analogue-to-digital converter if there is to be a resolution of 10 mV is:

A 500 bits
B 250 bits
C 9 bits
D 6 bits

17 Decide whether each of these statements is True (T) or False (F).

An analogue multiplexer has:
(i) An input of a number of analogue signals.
(ii) An output of digital versions of each analogue signal input.

Which option BEST describes the two statements?

A (i) T (ii) T
B (i) T (ii) F
C (i) F (ii) T
D (i) F (ii) F

18 Decide whether each of these statements is True (T) or False (F).

A cold junction compensator circuit is used with a thermocouple if it has:
(i) No cold junction.
(ii) A cold junction at the ambient temperature.

Which option BEST describes the two statements?

A (i) T (ii) T
B (i) T (ii) F
C (i) F (ii) T
D (i) F (ii) F

19 Decide whether each of these statements is True (T) or False (F).

A computer-based data display requires inputs which are:
(i) Voltages
(ii) Digital

Which option BEST describes the two statements?

A (i) T (ii) T
B (i) T (ii) F
C (i) F (ii) T
D (i) F (ii) F

20 Decide whether each of these statements is True (T) or False (F).

A data presentation elment which has an input which results in a pointer moving across a scale is an example of:
(i) An analogue form of display.
(ii) An indicator form of display.

Which option BEST describes the two statements?

A (i) T (ii) T
B (i) T (ii) F
C (i) F (ii) T
D (i) F (ii) F

21 Suggest sensors which could be used in the following situations:

(a) To monitor the rate at which water flows along a pipe and given an electrical signal related to the flow rate.
(b) To monitor the pressure in a pressurised air pipe, giving a visual display of the pressure.
(c) To monitor the displacement of a rod and give a voltage output.
(d) To monitor a rapidly changing temperature.

22 Suggest the type of signal processing element that might be used to:
(a) Transform an input of a resistance change into a voltage.
(b) Transform an input of an analogue voltage into a digital signal.
(c) Select one of a number of analogue signals for further processing.

3 Measurements

3.1 Measurement systems

In designing measurement systems there are a number of steps that need to be considered:

1 Identification of the *nature of the measurement* required.
For example, what is the variable to be measured, its nominal value, the range of values that might have to be measured, the accuracy required, the required speed of measurement, the reliability required, the environmental conditions under which the measurement is to be made, etc.

2 Identification of *possible sensors*.
This means taking into account such factors as their range, accuracy, linearity, speed of response, reliability, maintainability, life, power supply requirements, ruggedness, availability, cost. The sensor needs to fit the requirements arrived at in 1 and also be capable, with suitable signal processing, to give the required output for use in a control system in 3 and/or display in 4.

3 Selection of appropriate *signal processing*.
This element needs to take the output signal from the sensor and modify it in such a way as to enable it to drive the required display or be suitable for control of some device. For example, control applications might require a 4 to 20 mA current to drive an actuator.

4 Identification of the *required display*.
This means considering the form of display that is required. Is it to be an indicator or recorder? What is the purpose of the display?

This chapter is a consideration of some examples of measurement systems and the selection of the elements in such systems.

3.2 Examples of measurements

The following are examples of measurement problems and possible solutions. All of them can be related to measurements involved in cars and their modern electronic control systems.

3.2.1 Temperature

Consider the problem of the determination of temperature of a liquid in the range 0°C to 100°C where only rough accuracy is required. The situation might be the determination of the temperature of the cooling water for a car engine and its display as a pointer moving across a scale marked in different colours to indicate safe and unsafe operating temperatures. A solution might be to use a *thermistor* (see section 2.2.6) as a sensor. This is the commonly used solution with car engine coolant. The resistance change of the thermistor has to be converted into a voltage which can then be applied across a meter and so converted to a current through it and hence a reading on the meter related to the temperature.

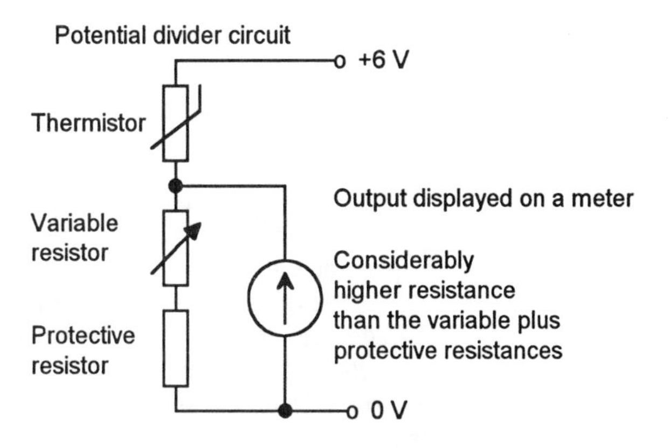

Figure 3.1 *Temperature measurement*

Figure 3.1 shows a possible solution involving a *potential divider circuit* (see section 2.4.1) to convert the resistance change into a voltage change. Suppose we use a 4.7 kΩ bead thermistor. This has a resistance of 4.7 kΩ at 25°C, 15.28 kΩ at 0°C and 0.33 kΩ at 100°C. The variable resistor might be 0 to 10 kΩ. It enables the sensitivity of the arrangement to be altered. However, if the variable resistor was set to zero resistance then, without a protective resistor, we could possibly have a large current passed through the thermistor. The *protective resistor* is there to prevent this occurring. The maximum power that the thermistor can withstand is specified as 250 mW. Thus, with a 6 V supply, the variable resistor set to zero resistance, the protective resistance of R, and the thermistor at 100°C, the current I through the thermistor is given by $V = IR$ as:

$$6 = I(0 + R + 330)$$

Thus:

$$I = \frac{6}{R + 330}$$

The power dissipated by the thermistor is $I^2 \times 330$ and so if we want this to be significantly below the maximum possible, say 100 mW, then we have:

$$0.100 = \left(\frac{6}{R + 330}\right)^2 \times 330$$

Hence R needs to be about 15 Ω.

When the temperature of the thermistor is 0°C its resistance is 15.28 kΩ. If we set the variable resistor as, say, 5 kΩ and the protective resistor as 15 Ω then the voltage output when the supply is 6 V is:

$$\text{output voltage} = \frac{5.015}{15.28 + 5.015} \times 6 = 1.48 \text{ V}$$

When the temperature rises to 100°C the output voltage becomes:

$$\text{output voltage} = \frac{5.015}{0.33 + 5.015} \times 6 = 5.63 \text{ V}$$

Thus, over the required temperature range, the voltage output varies from 1.48 V to 5.63 V. Thus a voltmeter to cover this range could be used to display the output.

The above calculation has assumed that the meter used to monitor this voltage has a much higher resistance than that of the variable plus protective resistances. If this is not the case than the loading effect (see section 1.3) of the meter resistance must be taken into account. This is because we have two resistances in parallel, the variable resistor plus the protective resistor and the meter resistance. Thus the effective resistance of that part of the circuit is changed by the presence of the meter. It is only if the meter has a very high resistance compared with that of the variable plus protective resistances that the effect of inserting the meter can be ignored.

In general, calibration of thermometers is by determining their response at temperatures which are specified as the standard values for freezing points and boiling points for pure materials. When a pure material changes state from liquid to solid or solid to vapour, the temperature remains constant at the change of state temperature (Figure 3.2) and provides an accurate reference temperature. Calibration of the thermistor thermometer at the 0°C temperature can be obtained by immersing the thermistor in a mixture of ice and water; calibration at the 100°C temperature by the thermistor being in the steam above boiling water. Alternatively, calibration at these and other temperatures within the range can be obtained by using a standard thermometer. Both the standard thermometer and the thermistor are placed in water which is slowly increased in temperature, the readings of the two being noted at a number of temperatures.

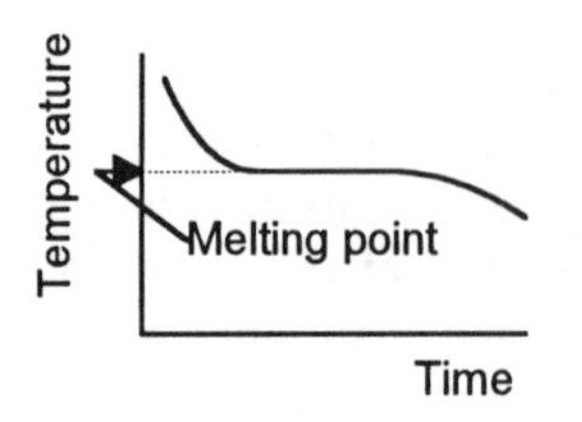

Figure 3.2 *Temperature-time curve of a material during solidifcation*

3.2.2 Pressure

Consider the problem of measuring an absolute pressure. It could be the manifold pressure in a car engine as part of the electronic control of engine power. The term *absolute pressure* is used for the pressure measured relative to zero pressure, the term *gauge pressure* being used if it is measured relative to the atmospheric pressure. A sensor that is used for such a purpose is a *diaphragm pressure gauge*. Figure 2.22 shows two forms of such a gauge.

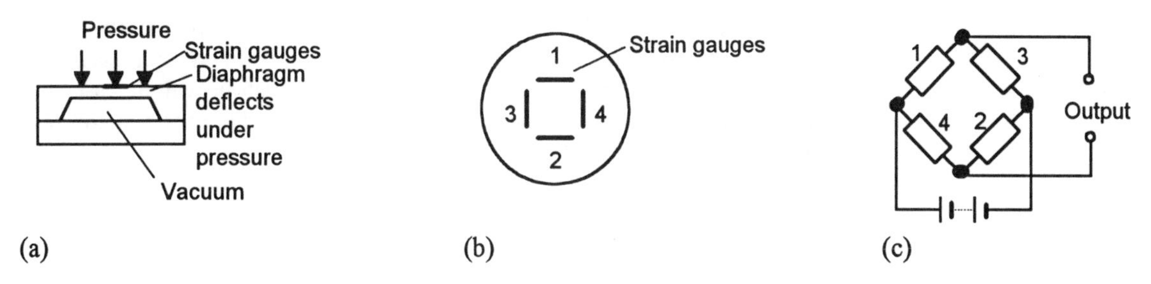

Figure 3.3 *Diaphragm pressure gauge: (a) basic form of sensor, (b) possible arrangement of strain gauges on diaphragm, (c) the Wheatstone bridge signal processor*

A typical specification of such a gauge might include:

Range: 0 to 200 kPa
Accuracy: nonlinearity and hysteresis ±0.1% of full scale reading
Repeatability: ±0.1% of full scale reading

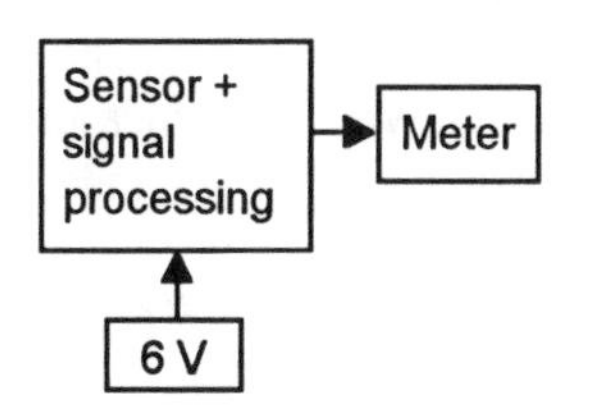

Figure 3.4 *Possible arrangement*

Figure 3.3(a) shows the basic form of a diaphragm pressure gauge which is often used in such circumstances. The diaphragm is made of silicon with the strain gauges diffused directly into its surface. Four strain gauges are used and so arranged that when two are in tension the other two are in compression (Figure 3.3(b). The four gauges are so connected as to form the arms of a Wheatstone bridge (Figure 3.3(c)). This gives temperature compensation since a change in temperature affects all the gauges equally (see section 2.4.2). Thus the output from the sensor with its signal processing is a voltage which is a measure of the pressure and can be displayed on a meter, possibly following some amplification. The arrangement using a commercial sensor with signal processor combined might thus be as shown in Figure 3.4.

Calibration of pressure gauges is usually with a *dead-weight pressure system*. Figure 3.5 shows the basic form of such a system.

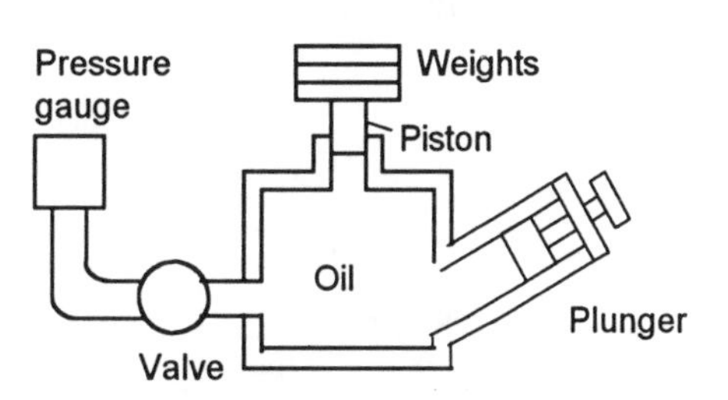

Figure 3.5 *Dead-weight pressure gauge calibration*

The calibration pressures are generated by adding standard weights to the piston tray, the pressure then being W/A, where W is the total weight of the piston, tray and standard weights and A the cross-sectional area of the piston. After the weights have been placed on the tray, the screw-driven plunger is screwed up to force the oil to lift the piston–weight assembly. Then the oil is under the pressure given by the piston–weight assembly since it is able to support that weight. By adding weights to the piston tray a gauge can be calibrated over its range.

3.2.3 Flow rate

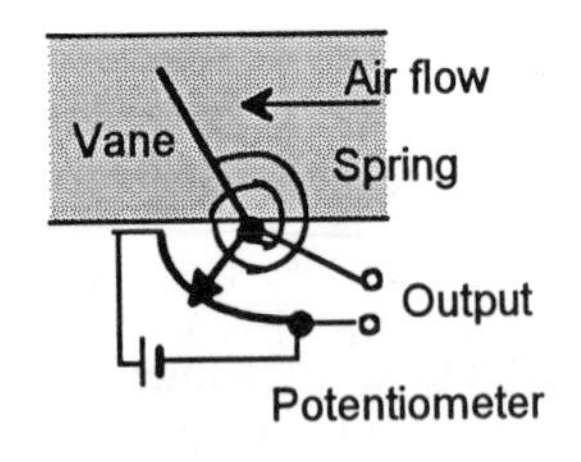

Figure 3.6 *Flow rate measurement*

Methods for the measurement of flow rate were discussed in section 2.2.4. However, consider the problem of measuring the flow rate of air for the infold of a car manifold in an electronic controlled engine. A simple, and cheap, measurement of the mass rate of flow of air is required with the output being an electrical signal which can be used for control purposes. One such method that is used is the *deflecting vane*. Figure 3.6 shows the basic principle. A vane is pivoted about one end and its movement restrained by a spring fixed at its pivot. For the vane to rotate it must move against the force exerted by the spring, the force increasing as the angle rotated increases. The air flow causes the vane to rotate, the angle deflected being a measure of the mass rate of flow. The angle through which the vane rotates can be turned into an electrical signal by its rotation moving a slider over a potentiometer wire. The output is thus a voltage which is a measure of the rate of flow. Such a flowmeter can be calibrated against a standard flowmeter, perhaps a Venturi meter.

An alternative which is used with cars is the *hot-wire anemometer*. This sensor consists of a platinum wire which is heated by an electrical current passing through it. The temperature of the wire will depend on the colling generated by the flow of air over the wire. Thus, since the electrical resistance of the wire will depend on its temperature, the resistance is a measure of the rate of flow of air over the heated wire. Figure 3.7(a) shows the basic form of such a sensor. The resistance change is transformed into a voltage change by incorporating the sensor as one of the arms of a Wheatstone bridge (Figure 3.7(b)).The bridge is balanced at zero rate of flow and then the out-of-balance voltage is a measure of the rate of flow. This voltage is fairly small and so has generally to be amplified.

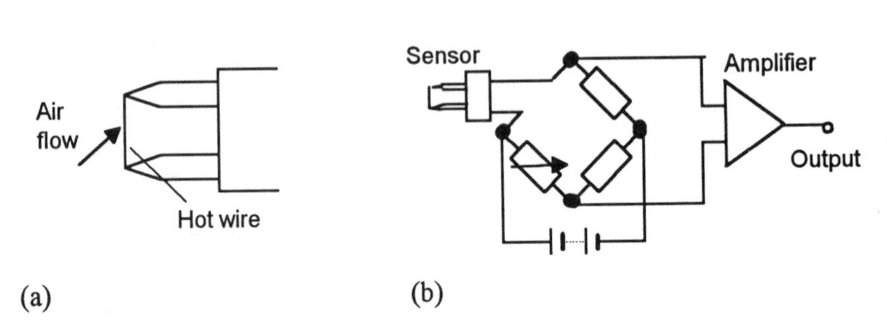

Figure 3.7 *Hot-wire anemometer*

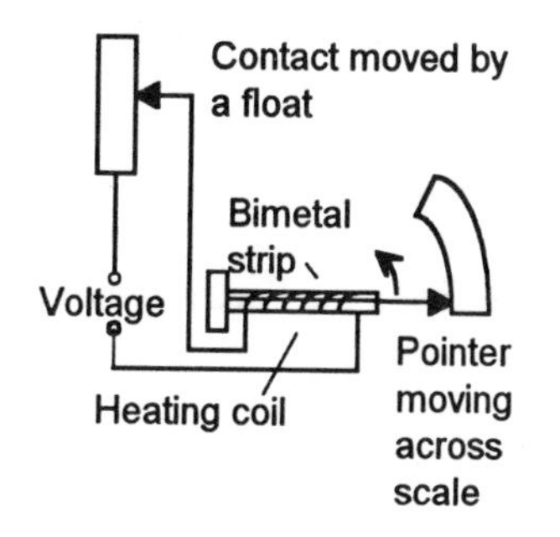

Figure 3.8 *Fuel level gauge*

3.2.4 Fluid level

Consider the problem of determining the level of petrol in the fuel tank of a car. A method that is used is the variable resistor (Figure 3.8). As the petrol level changes, so a float moves the slider of the variable resistor and hence there is a resistance change which depends on the height of petrol in the tank. For a constant voltage applied to the resistor, the current through it will thus be a measure of the fuel level. This current gives a display on a meter by passing through a heating coil wrapped round a bimetallic strip. As the current changes so the temperature of the coil changes and the bimetallic strip bends and gives a deflection related to the current and hence the fuel level.

3.2.5 Speed

Consider the problem of determining the speed of a car. The output from the measurement system is to be a display on an electrical meter and also might be used as an electrical signal in the electronic system used with a modern car. An assumption that is made is that the speed of a car is proportional to the rate at which the wheels are rotating. Thus the problem then becomes the measurement of the wheel speed.

This can be done by means of a tachogenerator (see section 2.2.2 and Figure 2.19). Figure 3.9(a) shows the basic form of such a generator. In practice there may be more poles on the magnet. The rotation of the 'drive' wheels is determined by the rotation of the drive shaft. This rotation can be used, via suitable gearing, to rotate the tachogenerator coil in the magnetic field (Figure 3.9(b)). The speed with which the drive shaft rotates determines the rate at which the coil rotates.

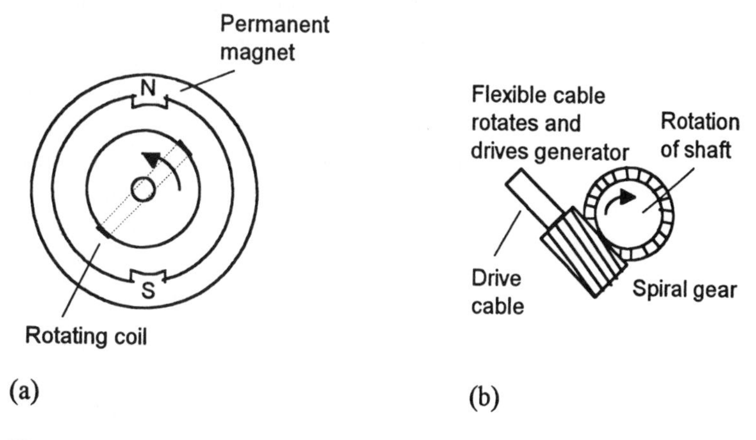

Figure 3.9 *(a) Tachogenerator principle, (b) coupling to drive shaft*

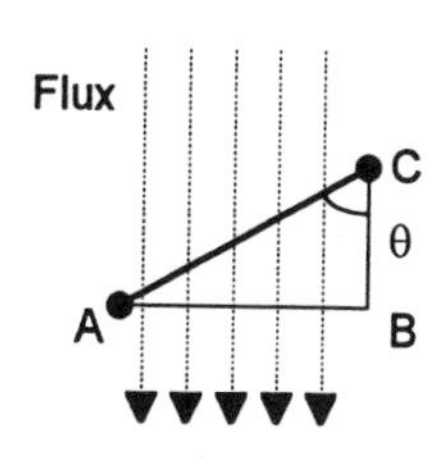

Figure 3.10 *Flux passing through the plane of the coil*

For the tachogenerator, the e.m.f. induced in the coil is proportional to the rate of change of magnetic flux linked by the coil (Faraday's law). As the coil rotates, the magnetic flux passing through the coil will change. In Figure 3.10, AC represents the plane of the coil when at an angle θ to the flux. The area through which the flux passes is thus AB. As the coil rotates to different angles, then AB changes. Since AB = AC sin θ, then the induced e.m.f. in the coil is given by

$$\text{e.m.f.} \propto \text{rate of change with time of } \sin\theta$$

For a coil rotating with constant angular speed ω the angle θ covered in a time t is given by $\theta = \omega t$. Thus the induced e.m.f. is given by

$$\text{e.m.f.} \propto \text{rate of change with time of } \sin\omega t$$

We can obtain the rate of change with time of sin ωt by differentiation.

$$\frac{\mathrm{d}(\sin\omega t)}{\mathrm{d}t} = \omega\cos\omega t$$

Thus the e.m.f. is given by

$$\text{e.m.f.} \propto \omega\cos\omega t$$

The e.m.f. is thus an alternating e.m.f. with an amplitude proportional to ω. Thus a meter displaying the 'size' of the alternating e.m.f., e.g. the root-mean-square value, gives a measure of the speed of rotation of the drive shaft and hence the speed of the car. The speedometer can be calibrated by comparison with a standard instrument.

3.2.6 Displacement

Consider the problem of determining the angular position of the accelerator pedal of a car. The output is to be an electrical signal which can be used as part of the electronic control system for the car. A simple way this can be done is to use the rotation of the pedal about its pivot to rotate the slider of a potentiometer over the resistance track. Figure 3.11 shows the basic form of such a measurement system. The output voltage from the potentiometer will depend on the angle of the sliding contact and hence the angular displacement of the accelerator pedal.

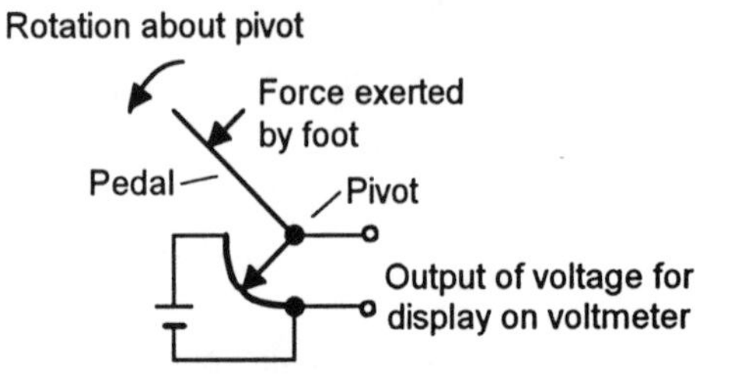

Figure 3.11 *Angular position measurement*

Problems

Questions 1 to 5 have four answer options: A, B, C and D. Choose the correct answer from the answer options.

1 Decide whether each of these statements is True (T) or False (F).

As part of the electronic control system for a car engine, a thermistor is to be used to monitor the air temperature. The signal processing circuit that could be used with the thermistor in order to give an electrical voltage output is:
(i) A Wheatstone bridge.
(ii) A potential divider circuit.

A (i) T (ii) T
B (i) T (ii) F
C (i) F (ii) T
D (i) F (ii) F

2 Decide whether each of these statements is True (T) or False (F).

It is proposed to monitor the exhaust temperature of a diesel engine by using a thermocouple. In order to give an output which is a few volts in size and which is independent of the temperature of the surrounding air temperature, the output from the thermocouple:
(i) Requires amplification.
(ii) Requires cold junction compensation.

A (i) T (ii) T
B (i) T (ii) F
C (i) F (ii) T
D (i) F (ii) F

3 Decide whether each of these statements is True (T) or False (F).

It is proposed to monitor the transmission oil pressure in a car by using a diaphragm pressure gauge with the movement of the diaphragm monitored by means of a linear variable differential transformer (LVDT).
(i) The output from the LVDT will be a resistance change which can be converted into a voltage change by a Wheatstone bridge.
(ii) The input to the LVDT is the displacement of the diaphragm.

A (i) T (ii) T
B (i) T (ii) F
C (i) F (ii) T
D (i) F (ii) F

4 Decide whether each of these statements is True (T) or False (F).

The signal processing needed for a system where the output from a thermocouple is to be fed into a microprocessor/computer includes:
(i) A digital-to-analogue converter.
(ii) Amplification.

A (i) T (ii) T
B (i) T (ii) F
C (i) F (ii) T
D (i) F (ii) F

5 Decide whether each of these statements is True (T) or False (F).

The signal processing needed for a system where the output from an optical encoder is to be fed into a microprocessor/computer includes:
(i) An analogue-to-digital converter.
(ii) A resistance-to-voltage converter.

A (i) T (ii) T
B (i) T (ii) F
C (i) F (ii) T
D (i) F (ii) F

6 A driverless vehicle is being designed for operation in a factory where it has to move along prescribed routes transporting materials between machines. Suggest a system that could be used to direct the vehicle along a route.

7 Identify the requirements of the measurement system and hence possible functional elements that could be used to form such a system for the measurement of:
(a) The production of an electrical signal when a package on a conveyor belt has reached a particular position.
(b) The air temperature for an electrical meter intended to indicate when the temperature drops below freezing point.
(c) The production of an electrical signal which can be displayed on a meter and indicate the height of water in a large storage tank.

4 Maintenance and testing

4.1 Maintenance and testing

If you own a car or motorcycle, then you will be involved with maintenance and testing, whether you carry out the procedures yourself or a garage does it for you. The function of maintenance is to keep the car/motorcycle in a serviceable condition to that it can continue to carry out its function of transporting you from one place to another. Maintenance is likely to take two forms. One form is *breakdown or corrective maintenance* in which repairs are only made when the car/motorcycle fails to work. Thus breakdown maintenance might be used with the exhaust with it only being replaced when it fails. The other form is *preventative maintenance* which involves anticipating failure and replacing or adjusting items before failure occurs. Preventative maintenance involves inspection and servicing. The inspection is intended to diagnose impending breakdown so that maintenance can prevent it. Servicing is an attempt to reduce the chance of breakdown occurring. Thus preventative maintenance might be used with the regular replacement of engine oil, whether it needs it or not at that time. Inspection of the brakes might be carried out to ascertain when new brake linings are going to be required so that they can be replaced before they wear out.

In carrying out maintenance, testing will be involved. Testing might involve checking the coolant level, brake fluid level, etc. and diagnosis tests in the case of faults to establish where the fault is.

This chapter is a brief consideration of the elements of maintenance and test procedures for engineering measurement systems and their functional elements. For details of the maintenance, test procedures/fault-finding checks required for a specific system or component, the manufacturer's manual for the system should be used.

4.2 Maintenance

In carrying out the maintenance of a measurement system, the most important aid is the *maintenance manual*. This includes such information as:

1 A description of the measurement system with an explanation of its use.
2 A specification of its performance.
3 Details of the system such as block diagrams illustrating how the elements are linked; photographs, drawings, exploded views, etc. giving the mechanical layout; circuit diagrams of individual elements; etc.
4 Preventative maintenance details, e.g. lubrication, replacement of parts, cleaning of parts and the frequency with which such tasks should be carried out.
5 Breakdown/corrective maintenance details, e.g. methods for dismantling, fault diagnosis procedures, test instruments, test instructions, safety precautions necessary to protect the service staff and precautions

to be observed to protect sensitive components. The fault diagnosis procedures might take the form of a series of programmed steps shown in block diagram form (Figure 4.1). With electrical systems the most commonly used test instruments are multirange meters, cathode ray oscilloscopes and signal generators to provide suitable test signals for injections into the system.

6 Spare parts list.

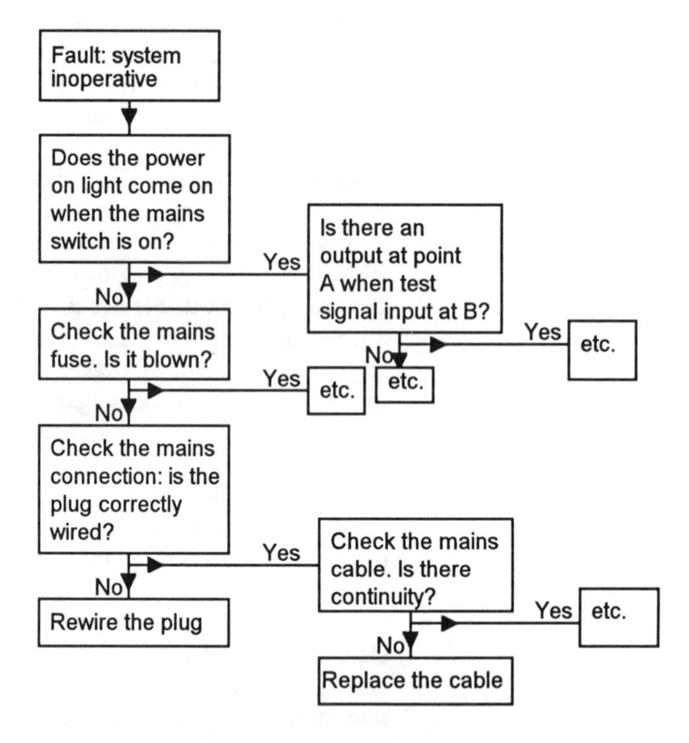

Figure 4.1 *Example of part of a fault diagnosis chart*

Maintenance can involve such activities as:

1 Inspection to determine where potential problems might occur or where problems have occurred. This might involve looking to see if wear has occurred or a liquid level is at the right level.
2 Adjustment, e.g. of contacts to prescribed separations or liquid levels to prescribed values.
3 Replacement, e.g. routine replacement of items as part of preventative maintenance and replacement of worn or defective parts.
4 Cleaning as part of preventative maintenance, e.g. of electrical contacts.
5 Calibration. For example, the calibration of an instrument might drift with time and so recalibration becomes necessary.

A record should be maintained of all maintenance activities, i.e. a maintenance activity log. With preventative maintenance this could take the form of a checklist with items being ticked as they are completed. Such a preventative maintenance record is necessary to ensure that such maintenance is carried out at the requisite times. The maintenance log should also include details of any adjustments made, recalibrations necessary or parts replaced. This can help in the diagnosis of future problems. The maintenance log can be part of the quality assurance procedures for a company (see section 1.4.1).

4.3 Fault diagnosis procedures

When presented with a faulty piece of equipment, in order to ascertain the cause of the fault a logical approach to fault diagnosis is required. This requires a systematic method to be adopted which essentially involves a set of sequential questions being posed, the answers to them determining the path followed through them. Figure 4.1 illustrated this with a block diagram giving some examples of such questions in a systematic sequence. Some of the questions can be based on:

1 *Physical inspection*
For example, inspection of a mains plug to check that all the wires are correctly connected to the right connectors. Visual inspection of a printed circuit board with the aid of a magnifying glass can be used to check for short-circuited components.

2 *Dynamic testing*
These checks are used to test the behaviour of parts of a system or components in a system by inputs being applied to test points and the outputs compared with what should occur. With a system consisting of a number of blocks connected in series (Figure 4.2), one method of locating the faulty block is to inject a suitable test signal into each of the test points in turn while monitoring the output and checking that the correct response is obtained. Thus with the test sequence indicated in Figure 4.2, a faulty output with test signal 1 would indicate a fault in block 4. Following the successful outcome of that test, then test signal 2 is used and the outcome checked. A faulty output would now indicate a fault in block 2, since block 3 has already been tested and found satisfactory. Then test signal 3 can be used. A faulty output would now indicate a fault in block 2. If there is no fault then test signal 4 can be used to establish whether the fault is block 1.

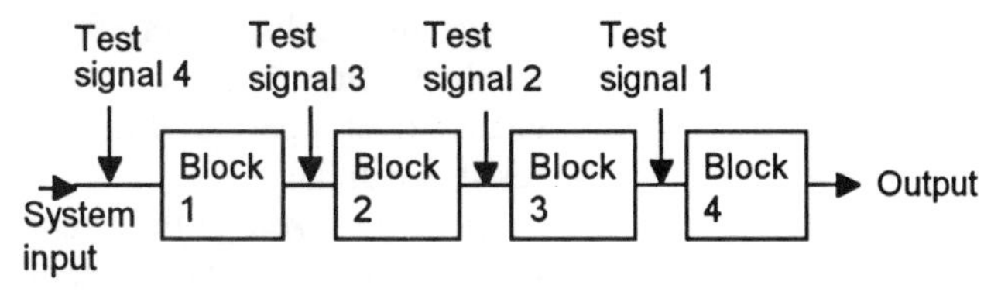

Figure 4.2 *Dynamic testing with a system having a number of series connected blocks*

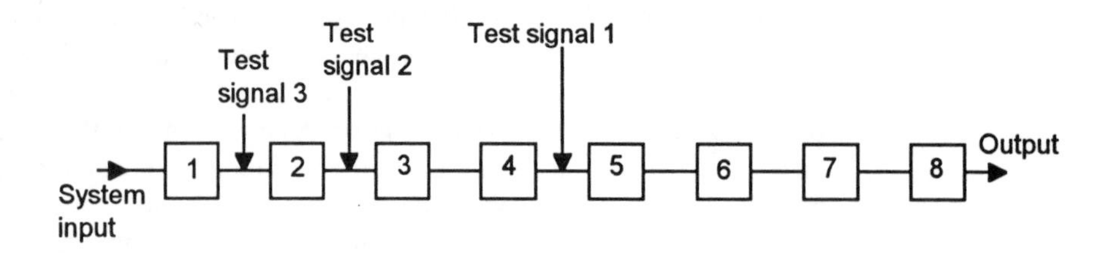

Figure 4.3 *Using the half-split technique of diagnosis*

An alternative to the systematic testing of a system block-by-block is the *half-split technique*. With a system having many blocks, this can be much faster at diagnosing a faulty block. With the sequential block-by-block testing, only one block is eliminated at each test, with the half-split technique, half of the blocks are eliminated at each test. The technique involves splitting the system in half. The output is then monitored when a test signal is applied at the halfway point. If there is a faulty output then the fault is in the second half sequence of blocks, if no fault then the fault is in the first half sequence. The faulty sequence of blocks is then tested by splitting it in half and applying a signal at its halfway point. The presence or otherwise of a faulty output then indicates which half the fault is in. The procedure can then be repeated until the faulty block is identified. Figure 4.3 illustrates this approach when the fault is in block 2.

3 *Replication checks*
This involves duplicating or replicating an activity and comparing the results. In the absence of faults it is assumed that the results should be the same. It could mean having duplicate systems and comparing the results given by the two, i.e. checking one instrument against another.

4 *Using reliability data*
The term *reliability* is used to describe the probability that an instrument, measurement system, subsystem within a measurement system, component, etc., will operate to an agreed level of performance. Manufacturer's data, the results of in-company testing, experience, etc., will yield information as to reliability, some systems/components being more reliable than others. Thus reliability data can indicate when systems/components should be replaced as part of preventative maintenance; also it can indicate which is most likely to be at fault when a system breaks down. For example, a filament lamp has a lower reliability than a carbon resistor. This means that there is a greater chance over a period of time that a filament lamp will fail than a carbon resistor will fail.

The *failure rate* is the average number of failures, per item, per unit time. Thus a high failure rate means a low reliability. If N items are tested for a time t with failed items being repaired and put back into service so that we always are testing N items, then if there are N_f failures, the failure rate is

$$\text{failure rate} = \frac{Nt}{N_f}$$

Table 4.1 gives some typical failure rates for electrical components and Table 4.2 failure rates for parts of measurement systems.

Table 4.1 Typical failure rates

Component	% failure rate per year
Carbon resistor	0.4
Wire wound resistor	0.09
Filament lamp	4.0
Soldered connection	0.009
Wrapped connection	0.0009
Silicon transistor	0.07
Integrated circuit	0.2

Table 4.2 Typical failure rates for parts of measurement systems

Component	% failure rate per year
Thermocouple in chemical environment	40
Resistance temperature element in laboratory	30
Bimetallic temp. indicator in industrial environment	130
Pen recorder in industrial environment	30
Bourdon pressure gauge in industrial environment	30

4.3.1 Common faults

The following are some of the commonly encountered tests and maintenance points that can occur with measurement systems:

1 *Sensors*
A test is to substitute a sensor with a new one and see what effect this has on the results given by the measurement system. If the results change then it is likely that the original sensor was faulty. If the results do not change then the sensor was not at fault and the fault is elsewhere in the system. Where a sensor is giving incorrect results it might be

because it is not correctly mounted or used under the conditions specified by the manufacturer's data sheet. In the case of electrical sensors their output can be directly measured and checked to see if the correct voltages/currents are given. They can also be checked to see whether there is electrical continuity in connecting wires.

2 *Switches and relays*
A common source of incorrect functioning of mechanical switches and relay is dirt and particles of waste material between the switch contacts. A voltmeter used across a switch should indicate the applied voltage when the contacts are open and very nearly zero when they are closed. If visual inspection of a relay discloses evidence of arcing or contact welding then it might function incorrectly and so should be replaced. If a relay fails to operate checks can be made to see if the correct voltage is across the relay coil and, if the correct voltage is present, that there is electrical continuity within the coil with an ohmmeter.

3 *Hydraulic and pneumatic systems*
A common cause of faults with hydraulic and pneumatic systems is dirt. Small particles of dirt can damage seals, block orifices, and cause moving parts to jam. Thus, as part of preventative maintenance, filters need to be regularly checked and cleaned. Also oil should be regularly checked and changed. Testing with hydraulic and pneumatic systems can involve the measurement of the pressure at a number of points in a system to check that the pressure is the right value. Leaks in hoses, pipes and fittings are common faults. Also, damage to seals can result in hydraulic and pneumatic cylinders leaking, beyond that which is normal, and result in a drop in system pressure.

Problems

Questions 1 to 4 have four answer options: A, B, C and D. Choose the correct answer from the answer options.

1 Decide whether each of these statements is True (T) or False (F).

The term preventative maintenance is used when:
(i) Systems are inspected to diagnose possible points of failure before failure occurs.
(ii) Systems are regularly maintained with such things as items being lubricated and cleaned.

A (i) T (ii) T
B (i) T (ii) F
C (i) F (ii) T
D (i) F (ii) F

Questions 2 to 4 refer to Figure 4.3, which shows a measurement system consisting of a number of blocks connected in series.

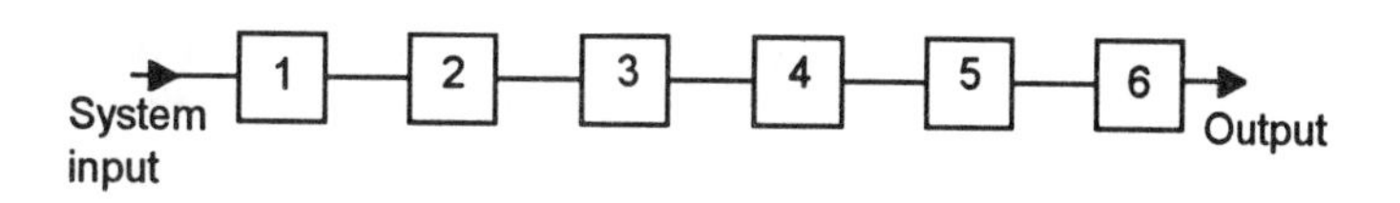

Figure 4.3 *Measurement system*

2 Decide whether each of these statements is True (T) or False (F).

When testing the system, there is no output from the system when a test signal is applied to the input to block 3. This can be because the faulty block is:
(i) Block 3.
(ii) Block 6.

A (i) T (ii) T
B (i) T (ii) F
C (i) F (ii) T
D (i) F (ii) F

3 Decide whether each of these statements is True (T) or False (F).

When testing the system, there is no output from the system when test signals are applied to the input to block 1 and then the input to block 5. This can be because the faulty block is:
(i) Block 2.
(ii) Block 6.

A (i) T (ii) T
B (i) T (ii) F
C (i) F (ii) T
D (i) F (ii) F

4 Decide whether each of these statements is True (T) or False (F).

When testing a faulty system, if there is a fault in block 5 there should be no output when a test signal is applied to the input to:
(i) Block 1.
(ii) Block 5.

A (i) T (ii) T
B (i) T (ii) F
C (i) F (ii) T
D (i) F (ii) F

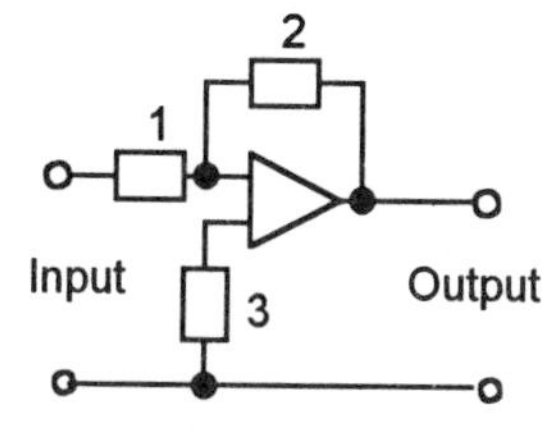

Figure 4.4 *Operational amplifier circuit*

5 Figure 4.4 shows an operational amplifier circuit. When all the resistors are working correctly and there is a test input of an alternating voltage of amplitude 150 mV, the output is an alternating voltage of amplitude 6 V. When there is a fault which makes resistor 1 open-circuit then there is no output. With the fault being resistor 2 open-circuit, the output is an alternating voltage which is deformed by severe clipping.

With the fault being resistor 3 open-circuit, the output is no alternating voltage but a d.c. voltage. Devise a flow diagram that can be used as a check list to identify faults in this circuit.

7 Describe the type of information that is likely to be contained in the maintenance manual for a system.

5 Control systems

5.1 Control systems

This chapter is a general discussion of the different types of control systems and their constituent functional elements. To illustrate the nature of control systems, consider some simple everyday and industrial systems.

1. You run a bath and adjust the temperature of the bath water by varying the amounts of water run in from the hot tap and the cold tap. This is an example of a control system with the variable being controlled being the water temperature.

2. You walk in a straight line along the pavement. This is an example of a control system with the variable being controlled being your position relative to a straight line.

3. You set the required temperature for a room by setting to the required temperature the room thermostat of a central heating system. This is an example of a control system with the variable being controlled being the room temperature.

4. You set the dials on the automatic clothes washing machine to indicate that 'whites' are being washed and the machine then goes through the complete washing cycle appropriate to that type of clothing. This is an example of a control system where a controlled sequence of events occurs.

5. The automatic clothes washing machine has a safety lock on the door so that the machine will not operate if the power is off and the door open. The control is of the condition which allows the machine to operate.

6. In a bottling plant the bottles are automatically filled to the required level. The variable being controlled is the liquid level in a bottle.

7. A packing machine is used to select the required number of items from a conveyor belt and pack them into a box. Control is being exercised such that items are being counted to obtain a specified number.

8. Packets of biscuits moving along a conveyor belt have their weights checked and those that are below the required minimum weight limit are automatically rejected. Control is being exercised over the weight.

9. A computer-numerical-control (CNC) machine tool is used to automatically machine a workpiece to the required shape, a controlled sequence of operations being carried out.

10 A belt is used to feed blanks to a pressing machine. As a blank reaches the machine, the belt is stopped, the blank positioned in the machine, the press activated to press the required shape, then the pressed item is ejected from the machine and the entire process repeated. A sequence of operations is being controlled with some operations controlled to occur only if certain conditions are met, e.g. activation of the press if there is a blank in place.

Considered as a whole, a *control system* can be thought of as a system which for some particular input or inputs is used to control its output to some particular value or give a particular sequence of events (Figure 5.1) or give an event if certain conditions are met.

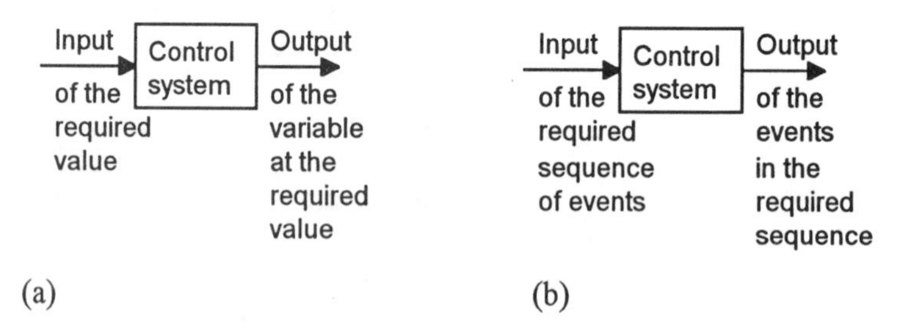

Figure 5.1 *Control systems used to: (a) control a variable to some required value, (b) control a sequence of events*

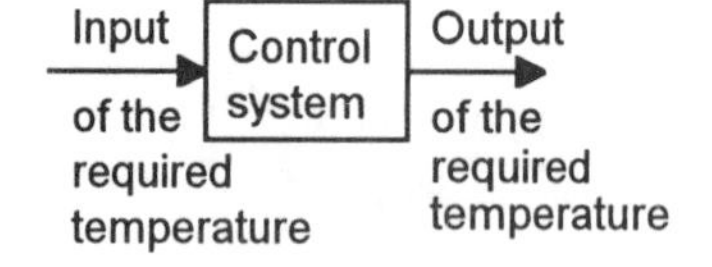

Figure 5.2 *Central heating system*

As an example of the type of control system described by Figure 5.1(a), a central heating control system has as its input the temperature required in the house and as its output the house at that temperature (Figure 5.2). The required temperature is set on the thermostat and the control system adjusts the heating furnace adjusts to produce that temperature. The control system is used to control a variable to some set value.

As an example of the type of control system described by Figure 5.1(b), a clothes washing machine has as its input a set of instructions as to the sequence of events required to wash the clothes, e.g. fill the drum with cold water, heat the water to 40°C, tumble the clothes for a period of time, empty the drum of water, etc.. The manufacturers of the machine have arranged a number of possible sequences which are selected by pressing a button or rotating a dial to select the appropriate sequence for the type of wash required. Thus the input is the information determining the required sequence and the output is the required sequence of events (Figure 4.3). The control system is used to control a sequence of events.

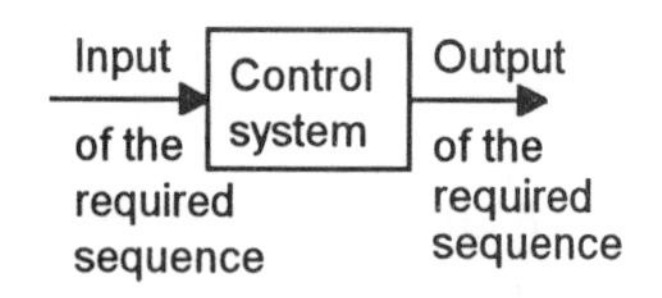

Figure 5.3 *Clothes washing machine system*

5.2 Open- and closed-loop control

Consider two alternative ways of heating a room to some required temperature. In the first instance there is an electric fire which has a selection switch which allows a 1 kW or a 2 kW heating element to be selected. The decision might be made that to obtain the required temperature it is only necessary to switch on the 1 kW element. The room will heat up and reach a temperature which is determined by the fact the 1 kW element is switched on. The temperature of the room is thus controlled by an initial decision and no further adjustments are made. This is an example of *open-loop control.* Figure 5.4 illustrates this. If there are changes in the conditions, perhaps someone opening a window, no adjustments are made to the heat output from the fire to compensate for the change. There is no information *fed back* to the fire to adjust it and maintain a constant temperature.

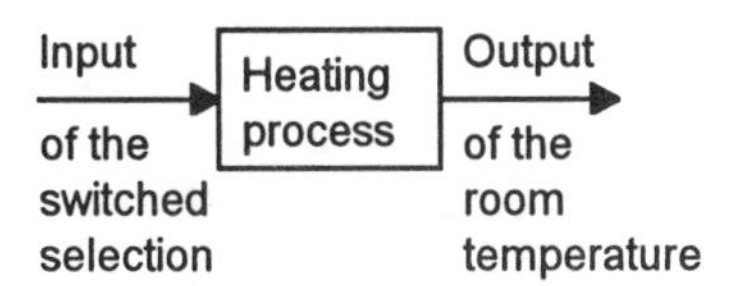

Figure 5.4 *The electric fire open-loop system*

Now consider the electric fire heating system with a difference. To obtain the required temperature, a person stands in the room with a thermometer and switches the 1 kW and 2 kW elements on or off, according to the difference between the actual room temperature and the required temperature in order to maintain the temperature of the room at the required temperature. These is a constant comparison of the actual and required temperatures. In this situation there is *feedback,* information being fed back from the output to modify the input to the system. Thus if a window is opened and there is a sudden cold blast of air, the feedback signal changes because the room temperature changes and so is fed back to modify the input to the system. This type of system is called *closed-loop*. The input to the heating process depends on the deviation of the actual temperature fed back from the output of the system from the required temperature initially set, the difference between them being determined by a comparison element. In this example, the person with the thermometer is the comparison element. Figure 5.5 illustrates this type of system.

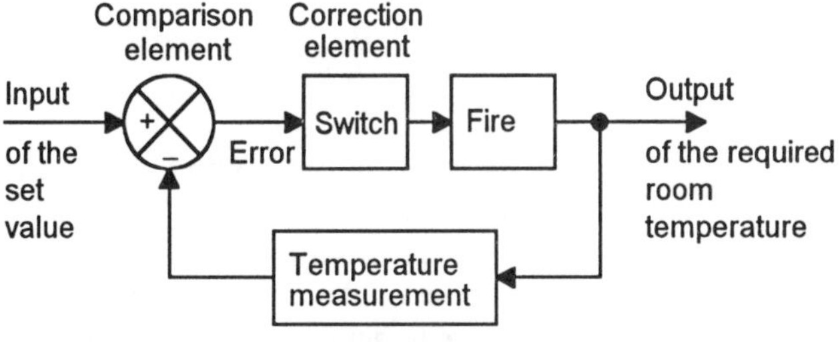

Figure 5.5 *The electric fire closed-loop system*

Note that the comparison element in the closed-loop control system is represented by a circular symbol with a + opposite the set value input and a – opposite the feedback signal. The circle represents a summing unit and what we have is the sum

+ set value – feedback value = error

This difference between the set value and feedback value, the so-called error, is the signal used to control the process. If there is a difference between the signals then the actual output is not the same as the desired output. When the actual output is the same as the required output then there is zero error. Because the feedback signal is subtracted from the set value signal, the system is said to have ***negative feedback***.

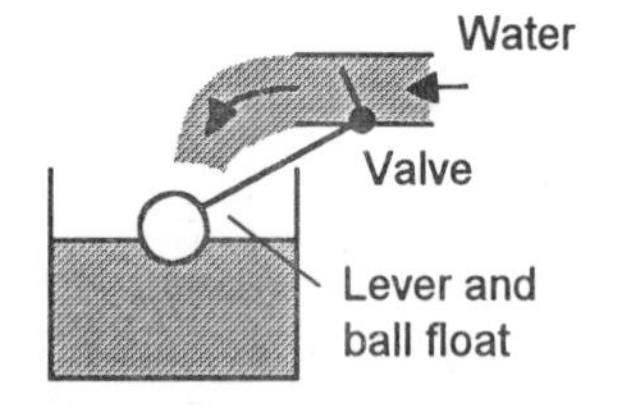

Figure 5.6 *Ball valve in a cistern*

Consider an example of a ball valve in a cistern used to control the height of the water (Figure 5.6). The set value for the height of the water in the cistern is determined by the initial setting of the pivot point of the lever and ball float to cut the water off in the valve. When the water level is below that required, the ball moves to a lower level and so the lever opens the valve to allow water into the tank. When the level is at the required level the ball moves the lever to a position which operates the valve to cut off the flow of water into the cistern. Figure 5.7 shows the system when represented as a block diagram.

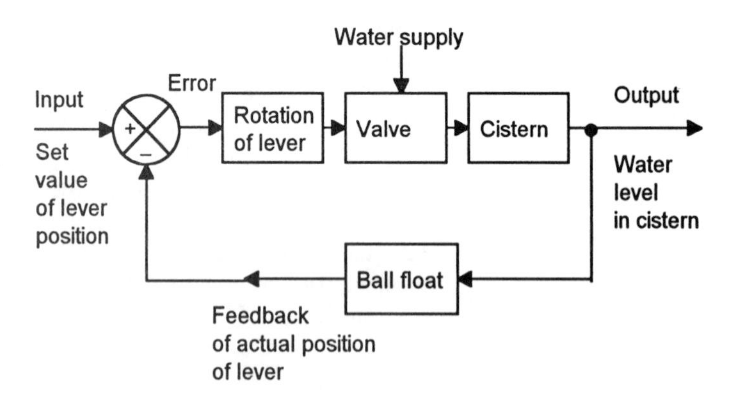

Figure 5.7 *Ball valve used to control water level in a cistern*

In an open-loop control system the output from the system has no effect on the input signal to the plant or process. The output is determined solely by the initial setting. In a closed-loop control system the output does have an effect on the input signal, modifying it to maintain an output signal at the required value. Open-loop systems have the advantage of being relatively simple and consequently cheap with generally good reliability. However, they are often inaccurate since there is no correction for errors in the output which might result from extraneous disturbances. Closed-loop systems have the advantage of being relatively accurate in matching the actual to the required values. They are, however, more complex and so more costly with a greater chance of breakdown as a consequence of the greater number of components.

5.2.1 Basic elements of a closed-loop system

Figure 5.8 shows the general form of a basic closed-loop system. It consists of a comparison element, signal processing to implement some control law, a correction element, the process being controlled and a measurement item to give a feedback signal.

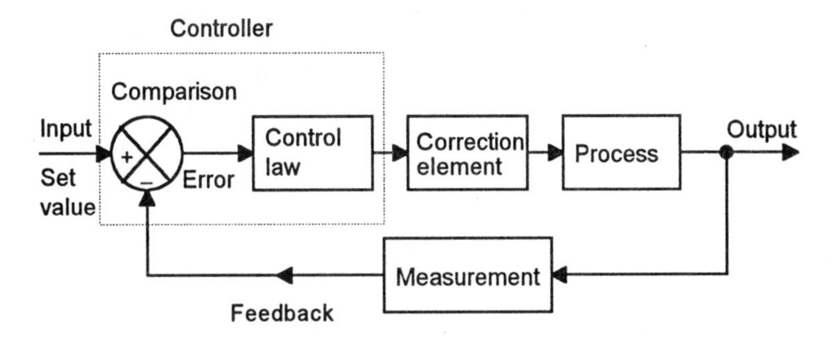

Figure 5.8 *Basic elements of a closed-loop control system*

The following are the functions of these elements:

1 *Comparison element*
This element compares the required value of the variable being controlled with the measured value of what is being achieved and produces an error signal.

$$\text{Error signal} = \text{reference value signal} - \text{measured value signal}$$

The symbol used for an element at which signals are summed is a segmented circle, inputs going into segments. The inputs are all added, hence the feedback input is marked as negative and the reference signal positive so that the sum gives the difference between the signals. A *feedback loop* is a means whereby a signal related to the actual condition being achieved is fed back to modify the input signal to a process. The feedback is said to be *negative feedback* when the signal which is fed back subtracts from the input value. It is negative feedback that is required to control a system. *Positive feedback* occurs when the signal fed back adds to the input signal.

2 *Control law implementation element*
The control law element determines what action to take when an error signal is received. The control law used by the element may be just to

supply a signal which switches on or off when there is an error, as in a room thermostat, or perhaps a signal which is proportional to the size of the error and thus proportionally opens or closes a valve according to the size of the error. Control plans may be *hard-wired systems* in which the control plan is permanently fixed by the way the elements are connected together or *programmable systems* where the control plan is stored within a memory unit and may be altered by reprogramming it. Programmable systems offer greater flexibility.

3 *Correction element*
The correction element or, as it is often called, the *final control element*, produces a change in the process which aims to correct or change the controlled condition. Thus it might be a switch which is used to switch on a heater and so increases the temperature of the process or perhaps a valve which opens and allows more liquid to enter the process. The term *actuator* is used for the element of a correction unit that provides the power to carry out the control action.

4 *Process element*
The process is what is being controlled. As illustrated in Figures 5.5 and 5.7, it might be a room in a house with its temperature being controlled or a tank of water with its level being controlled.

5 *Measurement element*
The measurement element produces a signal related to the variable condition of the process that is being controlled. It might be a temperature sensor with suitable signal processing or perhaps a ball float with a lever.

The term *control unit* is often used for the combination of the comparison element and the control law implementation element. Sometimes it is just used for the control law implementation element.

As a an example of a control system involving feedback, consider the motor system shown in Figure 5.9 for the control of the speed of rotation of the motor shaft. The input of the required speed value is by means of the setting of the position of the movable contact of the potentiometer. This determines what voltage is supplied to the comparison element, i.e. the differential amplifier, as indicative of the required speed of rotation. The differential amplifier produces an amplified output which is proportional to the difference between its two inputs. When there is no difference then the output is zero. The differential amplifier is thus used to both compare and implement the control law. The resulting control signal is then fed to a motor which adjusts the speed of the rotating shaft according to the size of the control signal. The speed of the rotating shaft is measured using a tachogenerator, this being connected to the rotating shaft by means of a pair of bevel gears. The signal from the tachogenerator gives the feedback signal which is then fed back to the differential amplifier.

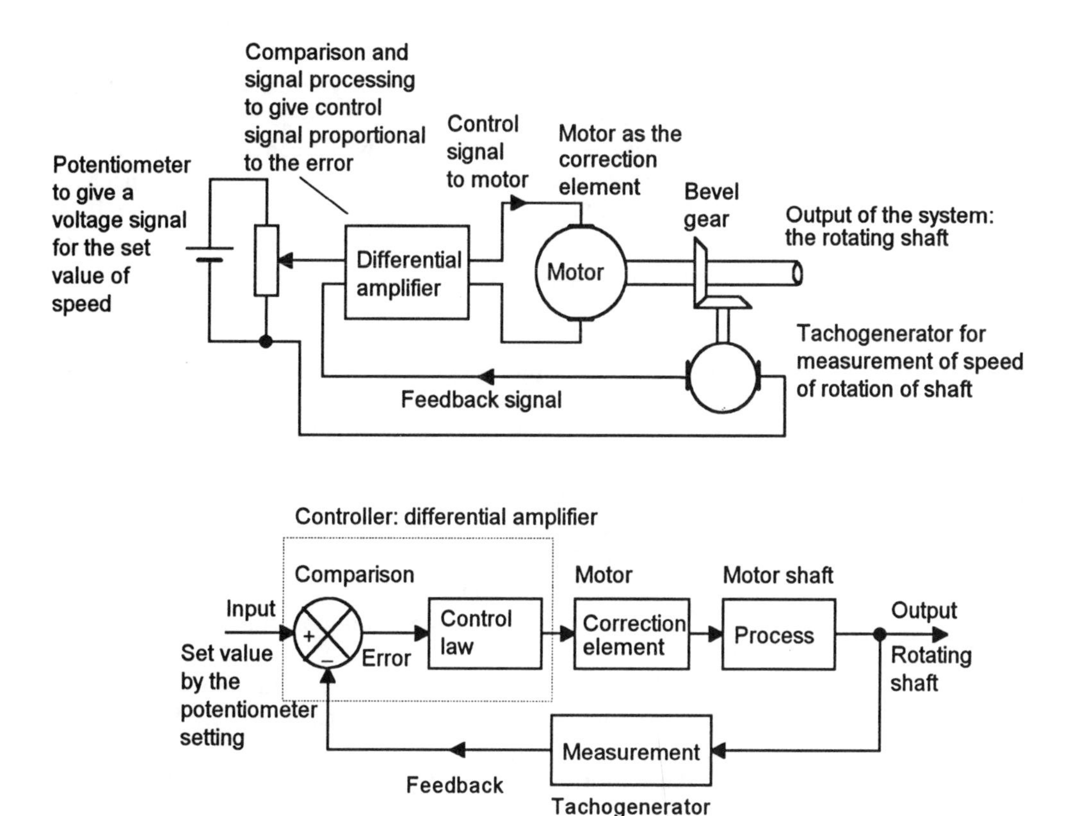

Figure 5.9 *Control of the speed of rotation of a shaft*

5.3 Sequential controllers

With the control system described above for the control of the shaft speed of a motor, the speed is being monitored all the time and the speed controlled all the time. Such a type of control system is often referred to as a *continuous control system*. There are, however, many control systems where the control is not exercised in this way but where the control is of a sequence of events so that one event follows another. Thus we might have the control that only after event 1 is complete that event 2 starts. When event 2 is complete then event 3 starts, etc. One action starts when the one before is complete. In other situations we might have one action occurring when a certain amount of time has elapsed and thus have a sequence of events being in a prescribed time sequence. In other cases we might have an action happening when a sensor starts it off and then perhaps continuing for a prescribed amount of time. This type of control system is called a *sequential control system*.

As a simple illustration of sequential control, consider the automatic kettle. When the kettle is switched on, the water heats up and continues heating until a sensor indicates that boiling is occurring. The kettle then automatically switches off. The heating element of the kettle is not continuously controlled but only given start and stop signals.

As another illustration of sequential control, consider the domestic washing machine. The machine has to carry out a number of operations in the correct sequence. The system operating sequence is called a ***program*** and there will be a number of programs which can be selected, the program depending on the type of clothes being washed in the machine. The sequence of instructions in each program is predefined and built into the controller used. Thus if the program for a white wash is selected the sequence of events that might be required is:

1 Start when switch pressed.
2 Open a valve and allow water into the machine drum for the prewash cycle.
3 When a sensor indicates that the water is to the required level, close the valve.
4 Switch on the heater and the tumbling action for the drum.
5 When the sensor indicates that the water temperature has reached 30°C, switch the heater off.
6 Tumble for a set time, 5 minutes.
7 Switch on a pump to empty the drum, continuing the tumbling during this operation.
8 Switch off the pump when a sensor indicates that it empty.
9 Open a valve and allow water into the machine drum for the main wash cycle.
10 When the sensor indicates that the water is to the required level, close the valve.
11 Switch on the heater and the tumbling action for the drum.
12 When the sensor indicates that the water temperature has reached 60°C, switch the heater off.
13 Tumble for a set time, 20 minutes.
14 Switch on a pump to empty the drum, continuing the tumbling action.
15 Switch off the pump when a sensor indicates that it is empty.
16 Open a valve and allow water into the machine drum for the rinse cycle.
17 When the sensor indicates that the water is to the required level, close the valve.
18 Tumble for a set time, 3 minutes.
19 Switch on a pump to empty the drum.
20 Switch off the pump when a sensor indicates that it is empty.
21 Open a valve and allow water into the machine drum for the second rinse.
22 When the sensor indicates that the water is to the required level, close the valve.
23 Tumble for a set time, 3 minutes.
24 Switch on a pump to empty the drum.
25 With the pump still on, spin for a set time, 4 minutes, at about 600 revolutions per second. Then cease spinning.
26 Continue pumping until the sensor indicates that the drum is empty, then switch off the pump.
27 Indicate on the machine control panel that the washing operation is ended.

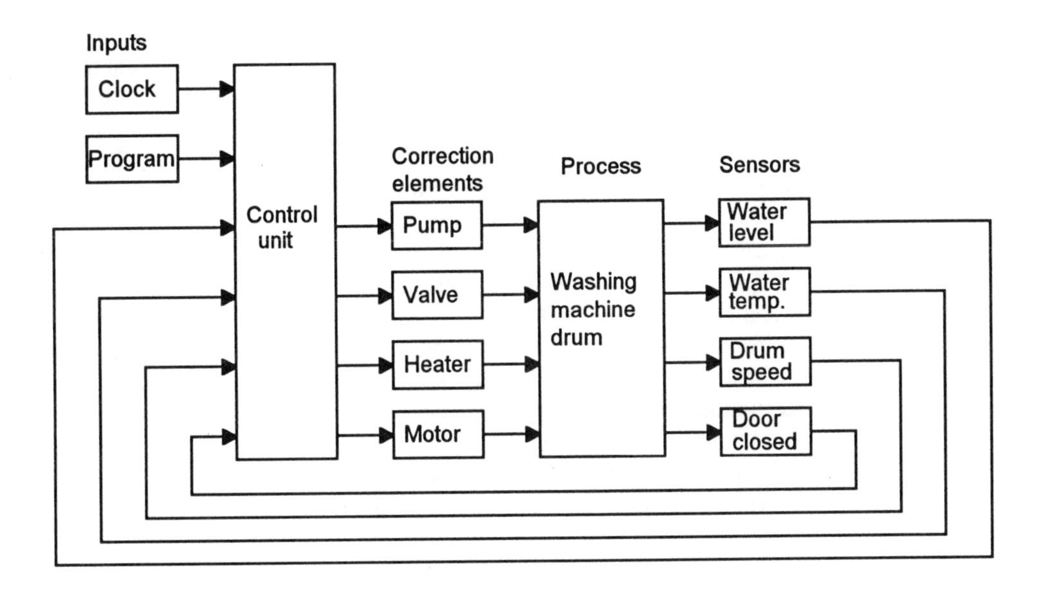

Figure 5.10 *The basic features of the washing machine*

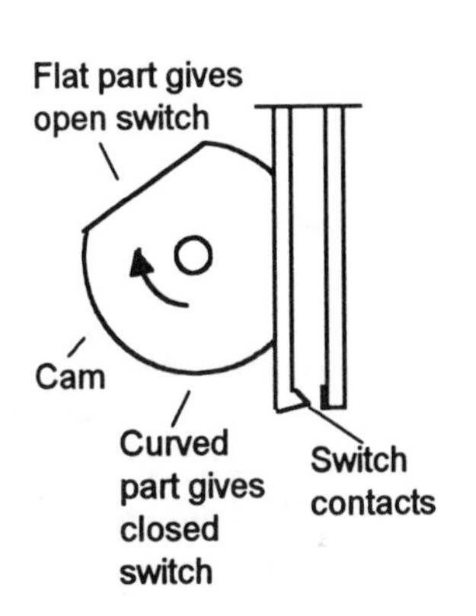

Figure 5.11 *A cam operated switch*

Figure 5.10 illustrates the basic features of the washing machine system. The controller has inputs from a clock, the program selector and the sensors. The outputs from the controller are signals to the water pump, the water valve, the heater and the drum motor. The sensors are for water level, the water temperature, the drum speed and whether the drum door is closed. The machine will not switch on when the drum door is open.

The washing machine requires operations to be switched on at a prescribed time for prescribed times. A system that has been used for this type of operation in the washing machine controller involves a set of cam-operated switches. Figure 5.11 shows the basic principle of one such switch. When the machine is switched on, a small electric motor rotates at a constant speed and so acts as a clock where the number of rotations is proportional to time. The motor rotates the controller cams so that each in turn operates electrical switches and so switches on circuits in the correct sequence. The contour of a cam determines the time at which it operates a switch and so the contours of the cams are the means by which the sequence and timing required for a program are specified and stored in the machine. The instructions used in a particular washing program are determined by the set of cams chosen. Thus the program selector determines which set of cams is used by the controller.

Thus when the washing machine is switched on, the clock motor starts to rotate the cams. The first cam that switches on in the pre-wash cycle is the one controlling an electrically operated valve which is opened to allow water to enter the drum when a current is supplied and switched off when it ceases. This valve allows cold water into the drum for a period of time determined by the profile of the cam used to operate its switch. However, since the requirement is a specific level of water in the washing machine

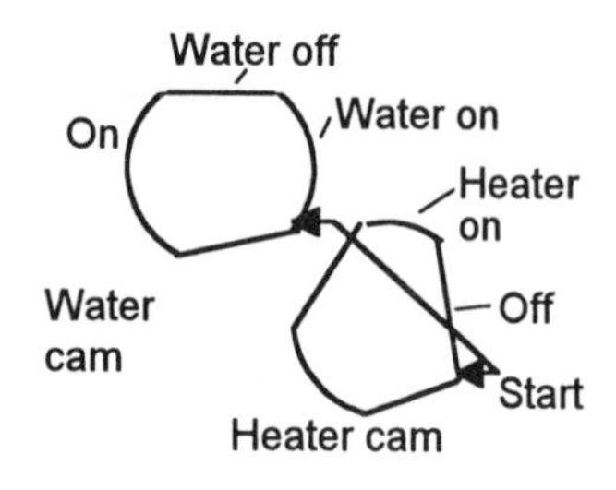

Figure 5.12 *Cams on a common shaft*

drum, there needs to be another mechanism which will stop the water going into the tank, during the permitted time, when it reaches the required level. In series with the cam-operated switch is a water level sensor which switches off the current to the valve when the water reaches the prescribed level. When the first cam switches off after its prescribed time, the next cam has just reached the point in its rotation when it can switch on the heater. Thus these two cams for these operations might look like those shown in Figure 5.12.

If you buy a new washing machine now, you are unlikely to find it operating using cams. The cam-operated controller has been replaced by a *programmable controller* involving a microprocessor. Figure 5.13 shows the basic structure of such a *microcontroller*. The microprocessor takes its instructions from the memory where the programs are stored. The input circuits and the output circuits provide protection for the microprocessor and make certain the voltages are at the right levels. They can also incorporate analogue-to-digital or digital-to-analogue converters.

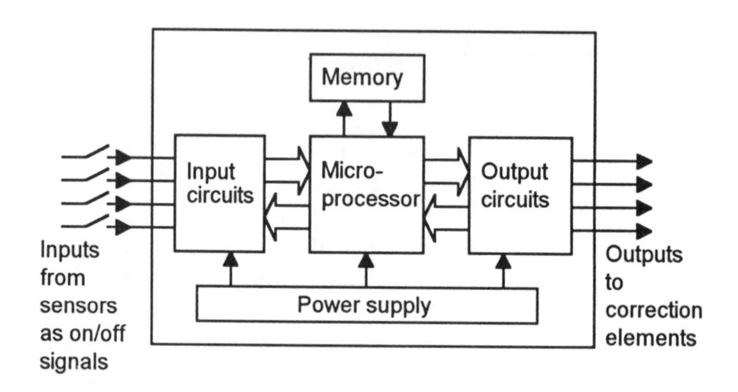

Figure 5.13 *Basic structure of a programmable controller*

Figure 5.14 shows the types of inputs and outputs that might be used in practice for a washing machine. Some of the inputs are analogue and so have to be converted to digital signals by an analogue-to-digital converter before passing to the input circuits. Other inputs are already digital and can be inputted through inputs which do not have an analogue-to-digital converter.

As an illustration of how a programmable controller can be used for a control system, consider a modern central heating system. The arrangement is likely to be of the form shown in Figure 5.15. The memory may store a number of programs, perhaps one for the night and one for the day. The inputs are the set temperature required for each program, the times for which each program is to run, and a signal for a temperature sensor. The output is to the heating system to switch on or off.

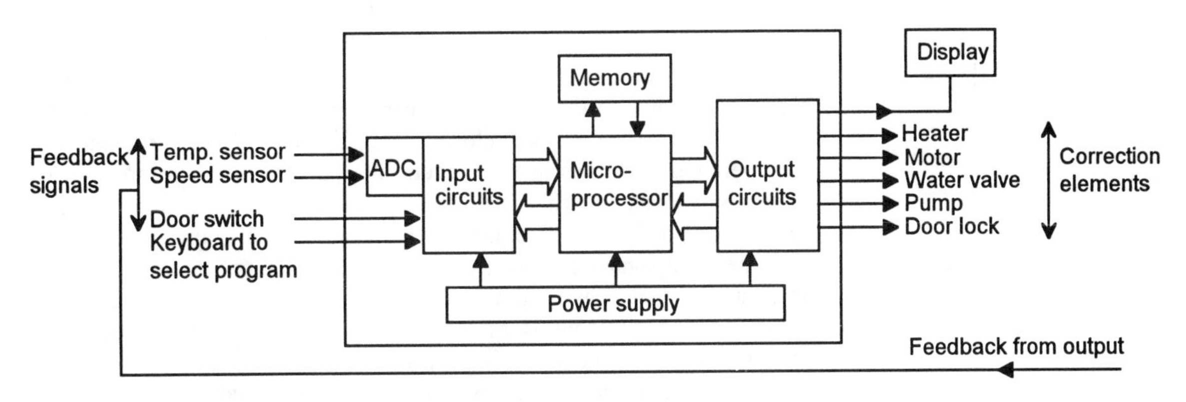

Figure 5.14 *The washing machine programmable control unit*

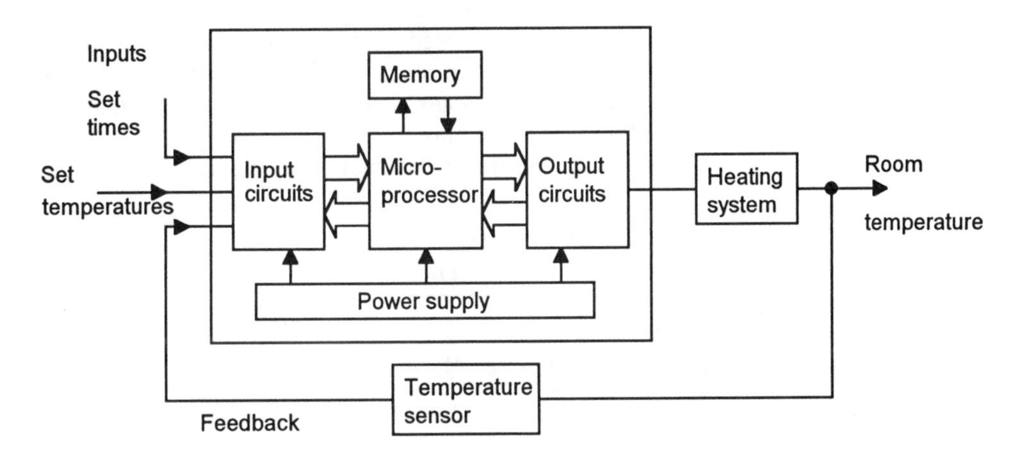

Figure 5.15 *Programmable controller used with a central heating system*

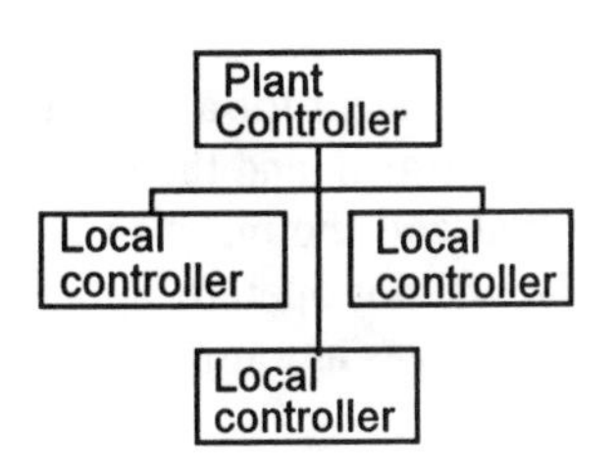

Figure 5.16 *Distributed control*

The term *direct digital control (DDC)* is used for systems where the comparison of the feedback with the set value and the determination of the signal or signals used to control the correction element or elements is exercised by a microprocessor, or computer, and such devices operate on digital signals.

The term *distributed control* is used for a system where microcontrollers are used to control local activities and themselves are controlled by another microcontroller or computer (Figure 5.16). Thus, for example, a chemical plant computer may be used to control local controllers controlling temperature, pressure, etc., at a number of points in the plant.

5.4 Logic gates

Many control systems involve *digital* signals where there are only two possible signal levels. These two levels may represent levels of on or off, open or closed, yes or no, true or false, +5 V or 0 V, etc. Such levels can be represented by the *binary* number system with the on, open, yes, true,

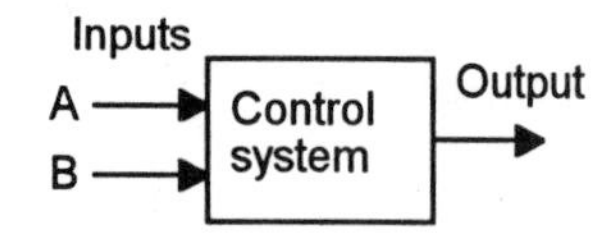

Figure 5.17 *Control system example*

+5 V levels being represented by 1 and the off, closed, no, false, 0 V levels by 0. This is the notation of *Boolean algebra*. With *digital control* the control is discontinuous. For example (Figure 5.17), we might have the water input valve to the domestic washing machine switched on if we have both the door to the machine closed, input signal A, and a particular time in the operating cycle has been reached, input signal B. There are two input signals which can be either yes or no signals and an output signal which can be a yes or no signal. The controller is programmed to give a yes output if both the input signals are yes and a no output when one or both of them are no. Thus if input *A* and input *B* are both 1 then there is an output of 1. If either input A or input B or both are 0 then the output is 0. Such an operation is said to be controlled by a *logic gate*, in this case an AND gate. Digital signals are processed by one or more logic gates. Microprocessors and computers operate in this way, being constructed from millions of such logic gates.

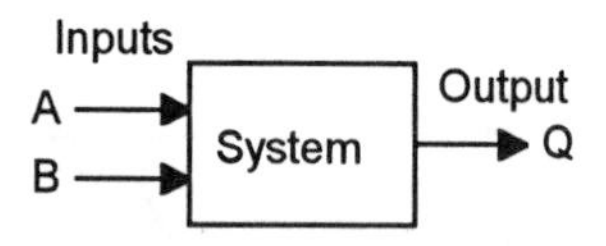

Figure 5.18 *Logic gate system*

The relationships between inputs to a logic gate and the outputs (Figure 5.18) can be tabulated in a form known as a *truth table*. Such a table specifies the relationships between the inputs and outputs. Thus for an AND gate with inputs *A* and *B* and a single output *Q*, we will have a 1 output when, and only when, $A = 1$ *and* $B = 1$. All other combinations of *A* and *B* will generate a 0 output. Thus the truth table is:

Inputs		Output
A	*B*	*Q*
0	0	0
0	1	0
1	0	0
1	1	1

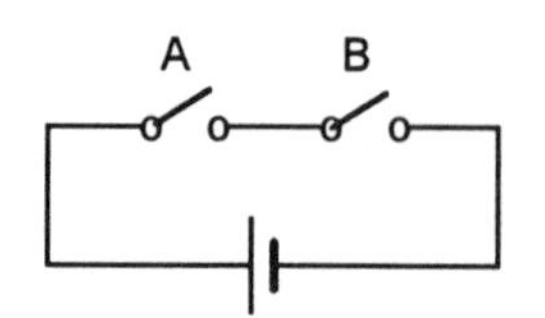

Figure 5.19 *An AND gate*

We can visualise the AND gate as an electrical circuit which has two switches, A and B, in series (Figure 5.19). Only when switch A *and* switch B are closed is there a current.

An example of an AND gate is an interlock control system for a machine tool such that if the safety guard is in place, giving a 1 signal, and the power is on, giving a 1 signal, then there can be a 1 output and the machine will operate. If either of the inputs is 0 then the machine will not operate.

An OR gate is a system which with inputs *A* and *B* gives an output of a 1 when *A or B* is 1. The following is the truth table:

Inputs		Output
A	*B*	*Q*
0	0	0
0	1	1
1	0	1
1	1	1

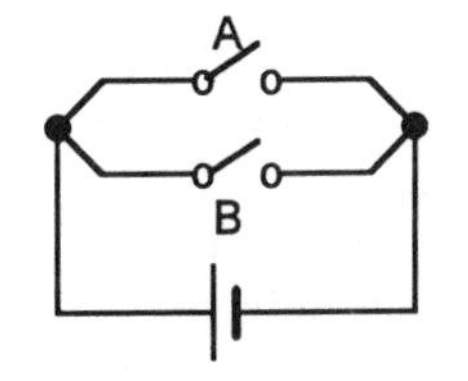

Figure 5.20 *An OR gate*

We can visualise the OR gate as an electrical circuit which has two switches in parallel (Figure 5.20). When switch *A or B* is closed then there is a current.

An example of an OR gate is a conveyor belt system transporting finished bottled products to packaging where an arm is required to deflect bottles off the belt if either the weight is not within certain tolerances or there is no cap on a bottle.

A NOT gate has just one input and one output, giving a 1 output when the input is 0 and a 0 output when the input is 1. The NOT gate gives an output which is the inversion of the input and is thus often called an *inverter*. The following is the truth table:

Input *A*	Output *Q*
0	1
1	0

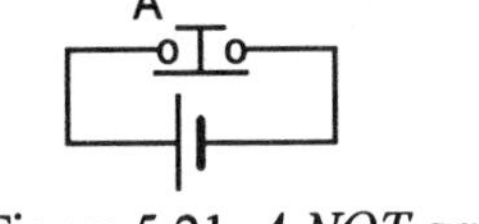

Figure 5.21 *A NOT gate*

We can visualise such a gate as being an electrical circuit (Figure 5.21) with a switch which is normally allowing current to pass but when pressed switches the current off.

An example of a situation where a NOT gate might be used is where a light has to come on when the light level falls below a set value. This might be a light which comes on at night. When there is an input there is not an output.

Gates can be combined to produce other relationships between inputs and outputs. Thus, for example, we might want to have no output when input A and input B are both on, and an output when either or both of them is off. We can generate such a relationship by combining an AND gate with a NOT gate (Figure 5.22). Such a combination is referred to as a NAND gate. The truth table is then:

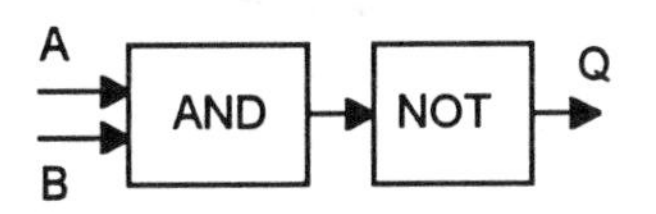

Figure 5.22 *Combining gates*

Inputs		Output from AND gate	Output from NOT gate
A	*B*		*Q*
0	0	0	1
0	1	0	1
1	0	0	1
1	1	1	0

Such a system might be used for a control situation where a light is to come on when the door bell is pressed or perhaps a signal is produced from a sensor which detects the presence of a person near the door and when it is not light, i.e. night time.

Likewise, other forms of gate can be produced by suitable combinations of gates. Thus if an OR gate is combined with a NOT gate we obtain a NOR gate, an exclusive OR (XOR) gate by a combination of OR and AND gates. So far gates have been represented as boxes with inputs and outputs. Standard symbols are, however, generally used for the different types of gate. Figure 5.23 indicates, for two inputs, the range of logic gates, their symbols and truth tables.

NOT

Input A	Output Q
0	1
1	0

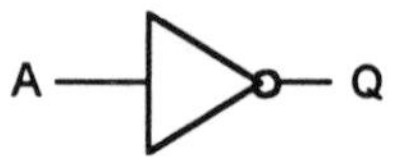

AND

Input A	Input B	Output Q
0	0	0
0	1	0
1	0	0
1	1	1

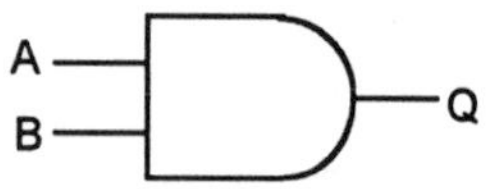

OR

Input A	Input B	Output Q
0	0	0
0	1	1
1	0	1
1	1	1

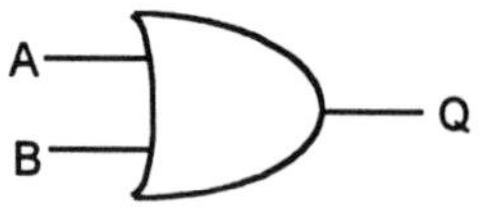

NAND

Input A	Input B	Output Q
0	0	1
0	1	1
1	0	1
1	1	0

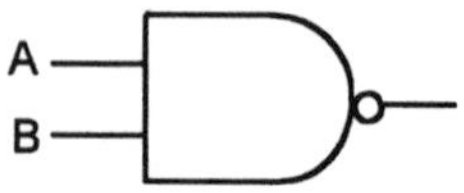

NOR

Input A	Input B	Output Q
0	0	1
0	1	0
1	0	0
1	1	0

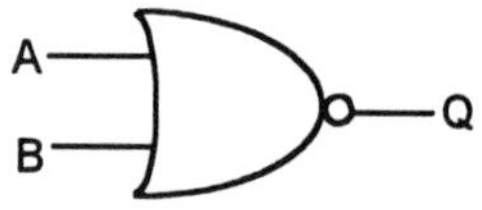

Figure 5.23 *Logic gates*

XOR

Input A	Input B	Output Q
0	0	0
0	1	1
1	0	1
1	1	0

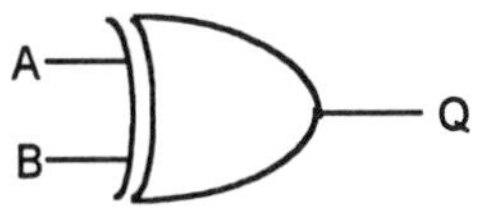

A
B
Q

XNOR

Input A	Input B	Output Q
0	0	1
0	1	0
1	0	0
1	1	1

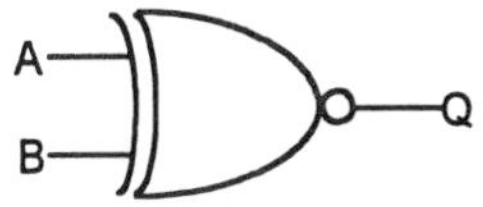

A
B
Q

Figure 5.23 *Logic gates (continued)*

Feedback
Input
Output

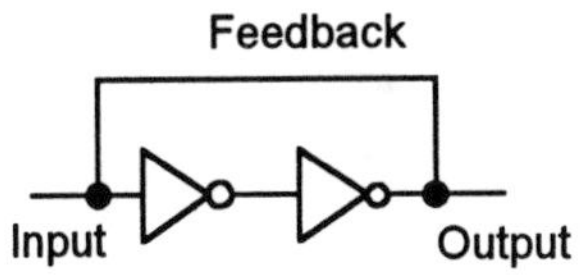

Figure 5.24 *An example of a sequential logic circuit. When input is 1, output is 1 and remains at that even if input is removed.*

5.4.1 Combinational and sequential logic circuits

If logic gates are combined in such a way that there is no feedback in the system, then the resulting circuit is said to be a *combinational logic circuit.* The examples quoted above are all combinational logic ones. If feedback is present, then there is the possibility that information can be stored in the circuit and the circuit is said to be a *sequential logic system* (Figure 5.24). With such a system the inputs to the circuit causes it to give an output which persists until there is some change of the inputs. When there is a change then the circuit settles down to a new stable state. The circuit thus follows a sequence of stable states, hence the term sequential.

5.4.2 Designing combinational control systems using logic gates

To determine what logic gate or combination of logic gates are required for a control system the following steps can be followed:

1 Draw a truth table with a column for each input and one for the output. For example if the output is to be controlled by three inputs the form of the truth table would be as follows:

Input 1	Input 2	Input 3	Output

2 Write in the table all the possible input combinations, writing a 1 for on and a 0 for off. Thus we might have an entries such as:

Input 1	Input 2	Input 3	Output
0	0	0	
1	0	0	

3 For each combination of inputs, work out whether the output is off or on, writing a 1 for on and a 0 for off. You might find some combinations of inputs where it does not matter whether the output is on or off, for these you might put 0/1 in the output column.

4 Write a statement, using terms such as AND, OR and NOT, saying when the output is on. This statement then provides the basis for the design of the system to give the correct output.

Logic gates are available as integrated circuit chips. For example, the integrated circuit 74LS21 has two 4-input AND gates in one 14 pin package (Figure 5.25). The integrated circuit 74LS08 has four 2-input AND gates in one 14 pin package. The integrated circuit 74LS32 has four 2-input OR gates in one 14 pin package.

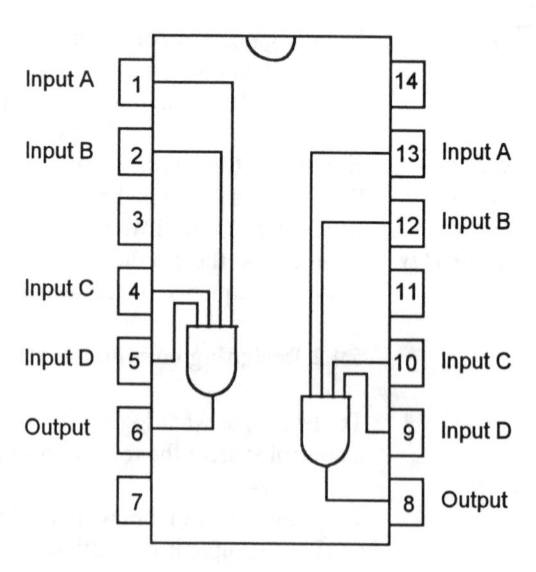

Figure 5.25 *The gates and pin connections for the 74LS21 integrated circuit package*

The microprocessor is a digital device based on the operation of logic gates. The advantage of using a microprocessor to implement logic functions is that they can be implemented by programming instructions to the microprocessor rather than hard wiring logic devices.

Example

What types of logic gates might be needed in the following control situations: (a) part of a chemical plant where an alarm is to be activated if the temperature falls below a certain level, (b) an automatic door is to open if a person approaches from either side?

(a) This requires a NOT gate in that the conditions required are:

Temperature	Alarm
Low (0)	On (1)
High (1)	Off (0)

(b) This requires an OR gate in that the conditions required are:

Person on side A	Person on side B	Door
Yes (1)	No (0)	Open (1)
No (0)	Yes (1)	Open (1)
No (0)	No (0)	Closed (0)
Yes (1)	Yes (1)	Open (1)

5.5 Which form of control?

Control systems can take a number of forms: open-loop control, closed-loop control, have outputs which are continuously monitored and controlled, have outputs which are controlled by just being switched on or off, sequential control where a sequence of actions are carried out, have a on/off output controlled by a number of on/off inputs. The following, being the examples of control systems discussed at the beginning of this chapter, illustrate the forms of control that might be used in some situations.

1 You run a bath and adjust the temperature of the bath water by varying the amounts of water run in from the hot tap and the cold tap. This is an example of a closed-loop control system (Figure 5.26) with the output being continuously monitored and changed.

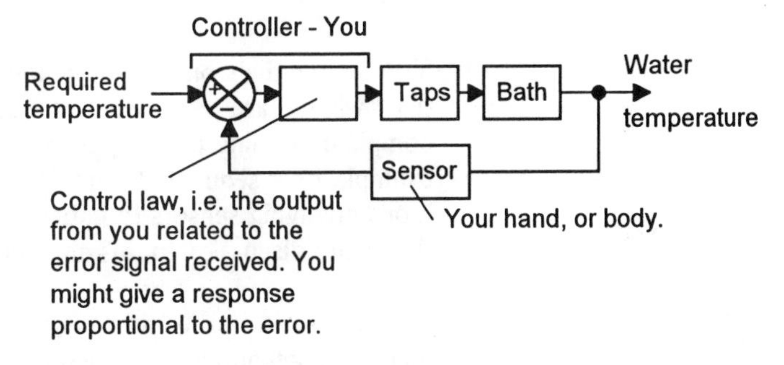

Figure 5.26 *The bath water control system*

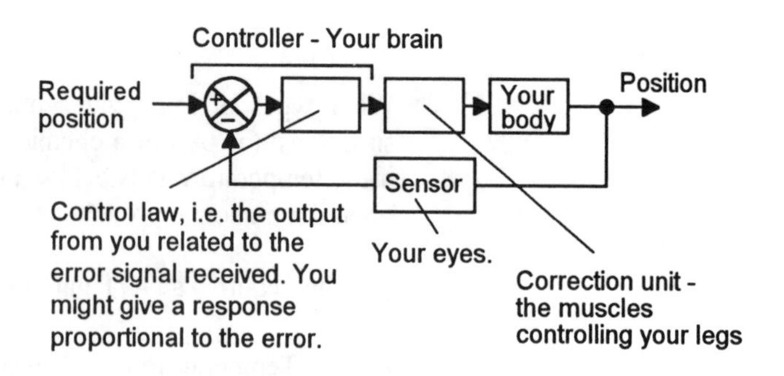

Figure 5.27 *The straight line walk control system*

2 You walk in a straight line along the pavement. This is an example of a closed-loop control system (Figure 5.27) with the feedback being provided by the eyes observing the person's position relative to the straight line and giving an error signal which is used to correct departures from the straight line. The output is continuously monitored and changed.

3 You set the required temperature for a room by setting to the required temperature the room thermostat of a central heating system. This is an example of a closed-loop control system (Figure 5.28). The heater is likely to be switched on if the room is too cold and off if it is at the right temperature or too warm. The output is thus on/off.

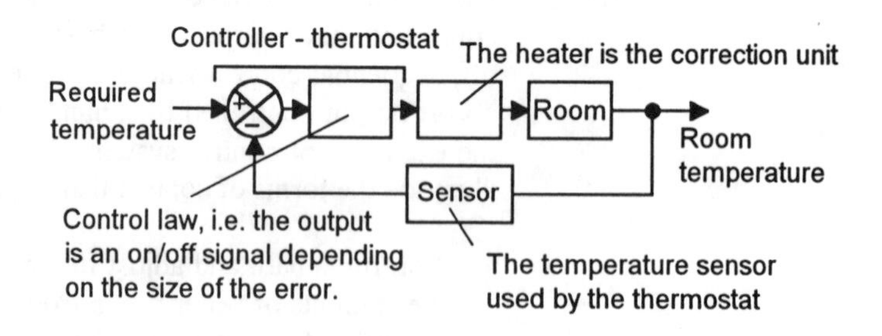

Figure 5.28 *Thermostat controlled heating system*

4 You set the dials on the automatic clothes washing machine to indicate that 'whites' are being washed and the machine then goes through the complete washing cycle appropriate to that type of clothing. This is an example of a sequential control system though there will be feedback loops involving sensors to indicate when some events should occur. See the earlier discussion in this chapter.

5 The automatic clothes washing machine has a safety lock on the door so that the machine will not operate if the power is off and the door open. This type of control is basically an AND logic gate (Figure 5.29).

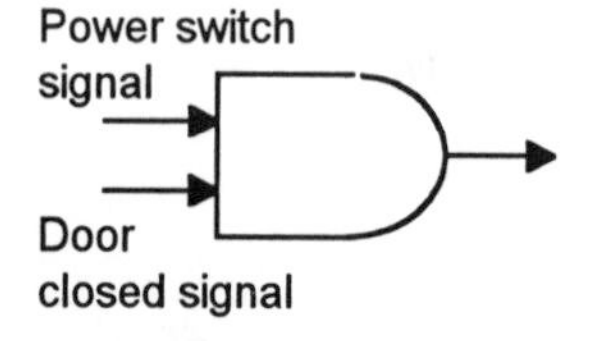

Figure 5.29 *AND gate system*

6 In a bottling plant the bottles are automatically filled to the required level. This is likely to be an example of an open-loop control system in that each bottle has a fixed volume of liquid as its input, regardless of whether the bottle already contains liquid or has a hole in it. There could be a closed-loop control system with the level of the liquid in a bottle being monitored by a sensor and information about the level fed back to adjust the liquid input.

7 A packing machine is used to select the required number of items from a conveyor belt and pack them into a box. The number of items is monitored and when the number reaches the required value an action is initiated. This might be regarded as one stage in a sequential control system.

8 Packets of biscuits moving along a conveyor belt have their weights checked and those that are below the required minimum weight are automatically rejected. This might be a simple logic gate system with the weight being monitored and the difference between the required weight and the actual weight being used to operate a NOT gate.

Input to gate	Output
Weight > minimum (1)	No rejection signal (0)
Weight < minimum (0)	Rejection signal (1)

9 A computer-numerical-control (CNC) machine tool is used to automatically machine a workpiece to the required shape, a controlled sequence of operations being carried out. This is a sequential control system with sensors being used to initiate and stop actions. It is, in some ways, rather like the washing machine example discussed earlier in this chapter. The part of the machining operation sequence involving cutting to a required shape can involve a feedback loop (Figure 5.30) with the actual position of the tool being compared with the position it should be and its position corrected if there is an error.

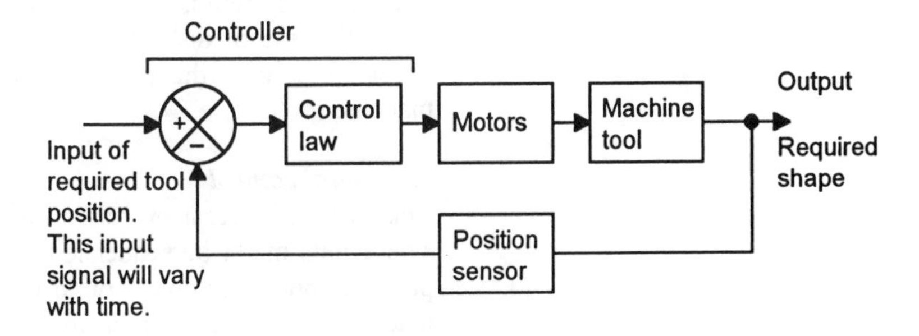

Figure 5.30 *Machining to the required shape*

This could be part of a distributed control system with a central computer being used to control local controllers at a number of machines and robots in order to give an automated production line.

10 A belt is used to feed blanks to a pressing machine. As a blank reaches the machine, the belt is stopped, the blank positioned in the machine, the press activated to press the required shape, then the pressed item is ejected from the machine and the entire process repeated. This is an example of sequential control with some operations controlled to only occur if certain conditions are met, e.g. activation of the press if there is a blank in place.

5.5.1 Classification of control systems

The following are some of the terms used to describe different types of control systems:

1 *Process control*
The term is used to describe the control of variables associated with a process in order to maintain them at some level. The process might, for example, be a central heating system for a house with the variable being controlled being temperature. The term process is here used for the ensemble of house and its contents, the heating boiler and radiator system. Another example is the control of water level in a cistern. The process here is the ensemble of water, cistern, water inlet and water outlet. The variable being controlled is water level.

2 *Servomechanisms*
The term servomechanisms is used for feedback control systems in which the controlled variable is a position or speed. An example of this is a motor controlled to maintain a constant speed of shaft rotation.

3 *Numerical control*
With such a system the information used as an input of the set values to the control system is stored in the form of digital codes on paper tape, magnetic tape or disk. This information can be used by the system in order to control the position, direction and speed of motion, of a machine tool.

4 *Sequential control*
This type of system exercises control over the sequencing of events. The events might be sequenced so that one cannot take place until the previous one is complete or to occur at particular times. The washing machine discusssed earlier in this chapter is an example of such a system.

Problems

Questions 1 to 13 have four answer options: A, B, C and D. Choose the correct answer from the answer options.

1 Decide whether each of these statement is True (T) or False (F).

An open-loop control system:
(i) Has negative feedback.
(ii) Responds to changes in conditions.

A (i) T (ii) T
B (i) T (ii) F
C (i) F (ii) T
D (i) F (ii) F

2 Decide whether each of these statement is True (T) or False (F).

A closed-loop control system:
(i) Has a measurement system which gives feedback of a signal which is a measure of the variable being controlled.
(ii) Has a controller which supplies a signal to a correction unit based on the difference between the set value and the fed back value for the variable being controlled.

A (i) T (ii) T
B (i) T (ii) F
C (i) F (ii) T
D (i) F (ii) F

3 Decide whether each of these statement is True (T) or False (F).

A sequential control system is one which can have:
(i) One event starting when the one before is complete.
(ii) One event starting when a certain amount of time has elapsed.

A (i) T (ii) T
B (i) T (ii) F
C (i) F (ii) T
D (i) F (ii) F

Questions 4 and 5 refer to the following forms of control systems:

A Open-loop
B Closed-loop
C Sequential
D Logic gate

4 Which type of control system is required for a domestic washing machine if it is only to switch on if the door is closed and the electrical power on?

5 Which type of control system is required if a room temperature is to be maintained constant regardless of any disturbances resulting from doors being opened?

Questions 6 to 9 refer to the following lines from a truth table:

Input A	Input B	Output Q
0	0	X
1	0	Y

6 For an OR gate we must have:

A $X = 0, Y = 0$
B $X = 1, Y = 1$
C $X = 0, Y = 1$
D $X = 1, Y = 0$

7 For an AND gate we must have:

A $X = 0, Y = 0$
B $X = 1, Y = 1$
C $X = 0, Y = 1$
D $X = 1, Y = 0$

8 For a NOR gate we must have:

A $X = 0, Y = 0$
B $X = 1, Y = 1$
C $X = 0, Y = 1$
D $X = 1, Y = 0$

9 For a NAND gate we must have:

A $X = 0, Y = 0$
B $X = 1, Y = 1$
C $X = 0, Y = 1$
D $X = 1, Y = 0$

10 Decide whether each of these statement is True (T) or False (F).

For the control system shown in Figure 5.31, the output Q will be 0 when:
(i) A, B, C and D are all 0.
(ii) A and B are 0, C and D are 1.

A (i) T (ii) T
B (i) T (ii) F
C (i) F (ii) T
D (i) F (ii) F

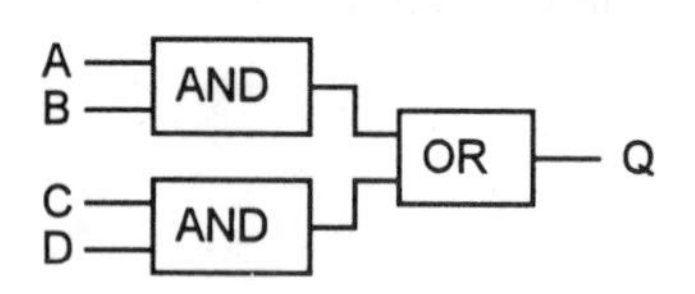

Figure 5.31 *Control system*

Questions 11 to 13 refer to the following logic gate systems:

A NOT
B AND
C OR
D NOR

11 Which gate system could be used for a red light to come on if the temperature of a heat treatment bath falls below a certain value?

12 Which gate system could be used with traffic lights if the light is to switch to go if a car is detected and a certain time has elapsed since it had switched to red?

13 Which gate system could be used if a burglar alarm is to sound if any one of the detectors is activated?

14 Suggest the possible form control systems might take for the following situations:
(a) Controlling the thickness of sheet steel produced by a rolling mill.
(b) A conveyor belt is to be used to transport packages from a loading machine to a pick-up area. The control system must start the belt when a package is loaded onto the belt, run the belt until the package arrives at the pick-up area, then stop the belt until the package is removed. Then the entire sequence can start again.
(c) Monitoring breathing in an intensive care unit, sounding an alarm if breathing stops.
(d) Controlling the amount of a chemical supplied by a hopper into sacks.
(e) Controlling the volume of water supplied to a tank in order to maintain a constant level.
(f) Controlling the illumination of the road in front of a car by switching on the lights.
(g) Controlling the temperature in a car by the driver manually selecting the heater controls, switching between them as necessary to obtain the required temperature.

15 Figure 5.32 shows two systems that might be used to control the temperature of a room. Explain how each operates.

16 Device a combinational logic gate system that could be used with a vending machine so that it dispenses either tea or coffee when the appropriate button is pressed and when money is inserted into the machine.

17 Devise a combinational logic gate system for use in a car so that an light will flash if when the key is turn to activate the ignition, the seat belts are not fastened.

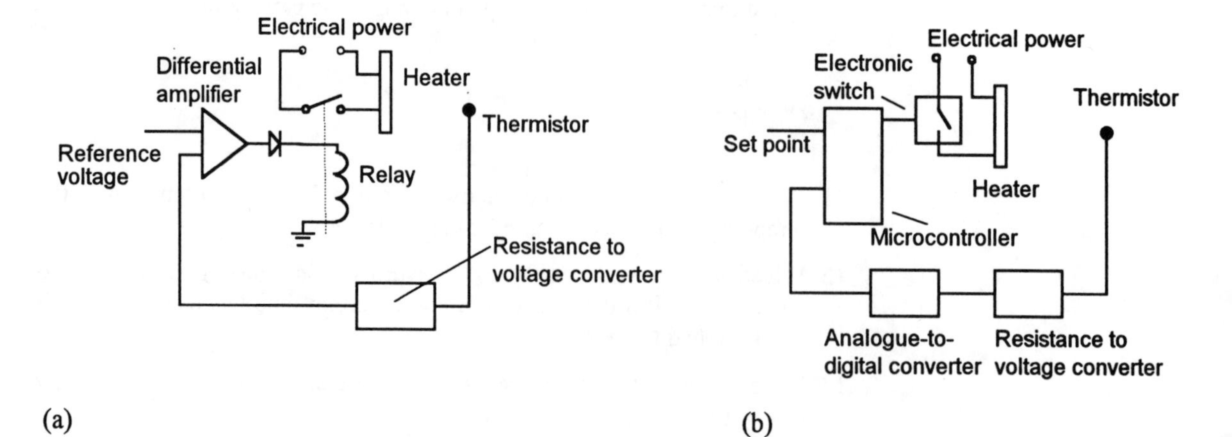

Figure 5.32 *Temperature-controlling systems*

6 Controllers

6.1 Controllers

This chapter is about the controllers used in control systems. The two forms of controllers considered are those concerned with closed-loop control when analogue, rather than digital, signals are involved and those involving the sequencing of events.

6.1.1 Closed-loop control

The controller in a *closed-loop control system* detects the error between the set value and the measured value and implements some method of control, converting the error into a control action designed to reduce the error. The error might arise as a result of some change in the conditions being controlled or because the set value is changed. The ways in which such controllers react to error changes are termed the *control laws*, or more often, the *control modes*. These are:

1. The *on/off mode* in which the controller is just a switch activated by the error signal, i.e. the control action is just on–off.
2. The *proportional mode* which produces a control action that is proportional to the error.
3. The *proportional plus derivative mode* (PD) which produces a control action that is composed of two terms, one which is proportional to the error and one that is proportional to the rate at which the error is changing.
4. The *proportional plus integral mode* (PI) which produces a control action that is composed of two terms, one which is proportional to the error and one that is proportional to the integral of the error.
5. The *proportional plus integral plus derivative modes* (PID), or the *three term mode*. This produces a control action that is composed of three terms: one which is proportional to the error, one which is proportional to the integral of the error and one which is proportional to the rate at which the error is changing.

These five modes of control are discussed in the following sections of the chapter.

In any control system with feedback the system cannot respond instantly to any change and thus there are delays while the system takes time to accommodate the change. Such delays are referred to as *lags*. For example, in the control of the temperature in a room by means of a central heating system, if a window is suddenly opened and the temperature drops or the thermostat is suddenly set to a new value, a lag will occur before the control system responds, switches on the heater and gets the temperature back to its set value.

6.1.2 Sequential control

Sequential control is the sequencing of events, e.g. the switching on or off of pumps, heaters, etc. in a washing machine, and in this chapter the controller discussed for this purpose is the programmable logic controller (PLC).

6.2 On–off mode

An example of the *on–off mode*, or, as it is sometimes termed, the *two-step mode*, of control is the control system involving a heater which is switched on or off according to the temperature. If the room temperature is above the required temperature then the switch is in an off position and the heater is off. If the room temperature falls below the required temperature then the switch moves into an on position and the heater is switched fully on. The error signal is thus used by the controller to give a signal which switches the heater on or off.

Figure 6.1 shows the type of response that might occur. When the system is initially switched on, there is an error signal because the room temperature is below the set temperature. Thus the heater is switched on and the temperature rises. When the set value is reached, the error signal becomes zero but because of the time taken for the system to react, the temperature overshoots the set value. This time includes the time taken by the control system to react but also the time taken for the heater to cool down after being switched off. When the room temperature drops below the required level time elapses before the control system responds, switches the heater on and the heater begins to have an effect on the room temperature. In the meantime the temperature has fallen even more. The temperature thus undershoots the set value. The result is that the room temperature oscillates above and below the required temperature. This effect is termed *hunting*.

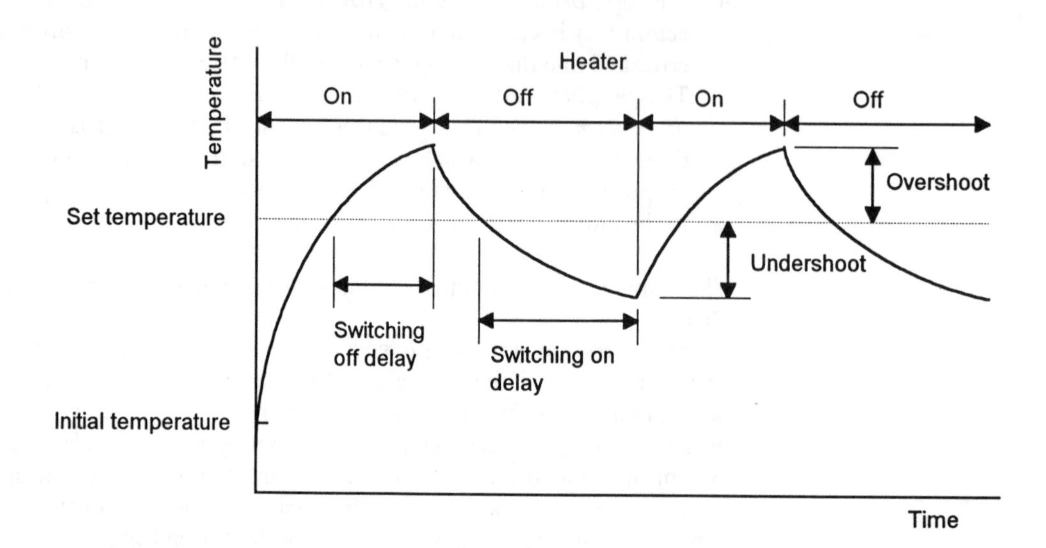

Figure 6.1 *Temperature changes with an on–off mode*

The time taken for the system to respond to a change in temperature is affected by the size of the room, the size of the heater and the sensitivity of the control system. On–off control action tends to be used where changes in the variable take place very slowly and so gives oscillations about the set value with a long periodic time. On–off control is thus not very precise in maintaining the set value, but it does involve simple devices and is thus fairly cheap.

6.2.1 On–off devices

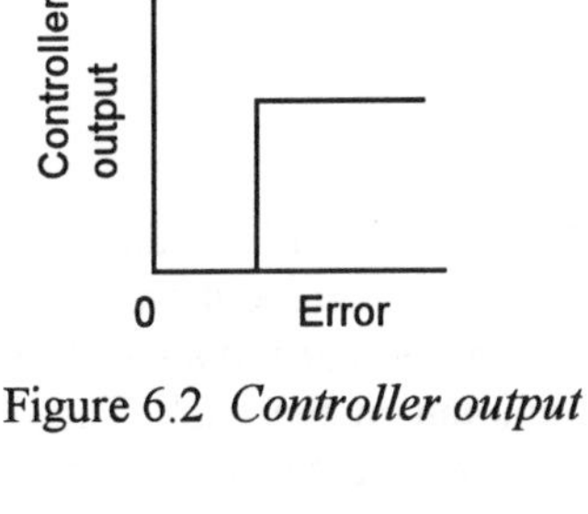

Figure 6.2 *Controller output*

Figure 6.2 shows how the output signal from an on-off controller is related to the error signal. An on–off device is essentially a switch which is activated by the error signal.

A widely used form of such a controller is a *relay*. Figure 6.3 shows the basic form of an electromagnetic relay. A small current at a low voltage applied to the solenoid produces a magnetic field and so an electromagnet. When the current is high enough, the electromagnet attracts the armature towards the pole piece and in doing so operates the relay contacts. A much larger current can then be switched on. When the current through the solenoid drops below the critical level, the springy nature of the strip on which the contacts are mounted pushes the armature back to the off position. Thus if the error signal is applied to the relay, it trips on when the error reaches a certain size and can then be used to switch on a much larger current in a correction element such as a heater or a motor.

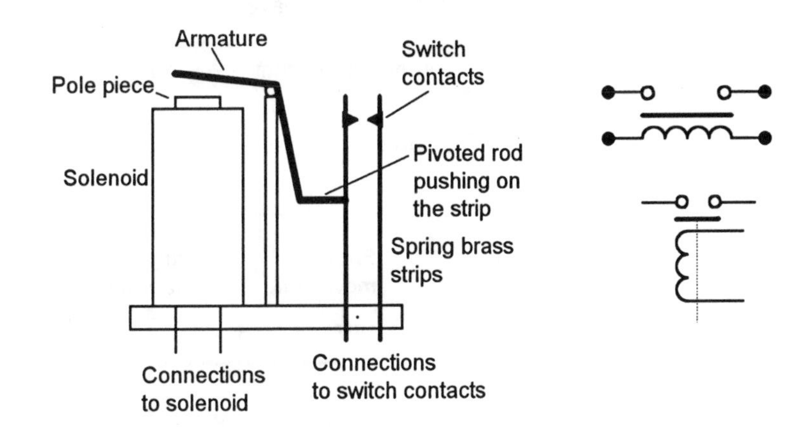

Figure 6.3 *The electromagnetic relay and symbols used to represent it*

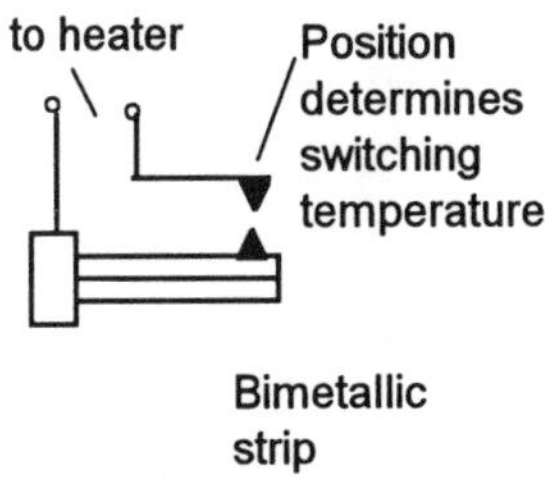

Connections to heater

Position determines switching temperature

Bimetallic strip

Figure 6.4 *Bimetallic thermostat*

An example of an on–off controller is a bimetallic switch thermostat. Figure 6.4 shows the basic principle. As the temperature decreases so the bimetallic strip bends towards the contact, making contact when the temperature falls to some particular value. This then switches on the heater. When the temperature rises sufficiently, the bimetallic strip moves away from making contact and the heater is switched off.

6.3 Proportional control

With the on–off method of control, the controller output is either an on or an off signal, the size of the on signal being independent of the size of the error (see Figure 6.2). With *proportional control* the size of the controller output is proportional to the size of the error (Figure 6.5), i.e. the controller input. Thus we have: controller output $\propto$ controller input. We can write this as:

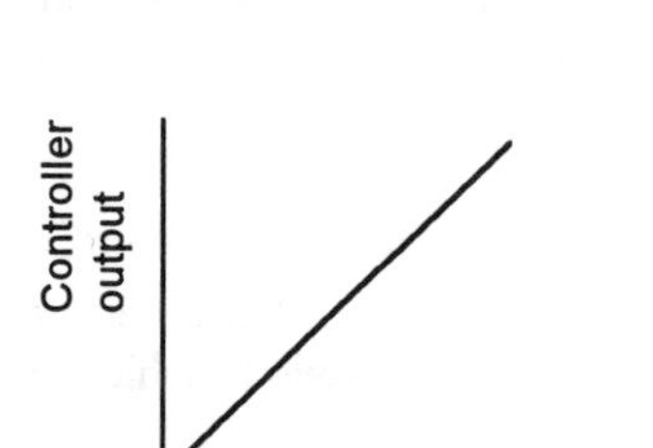

Figure 6.5 *Controller output*

$$\text{controller output} = K_p \times \text{controller input}$$

where K_p is a constant called the gain:

$$\text{gain } K_p = \frac{\text{controller output}}{\text{controller input}}$$

This means the correction element of the control system will have an input of a signal which is proportional to the size of the correction required.

To illustrate the above, consider a person exercising control over his/her bank balance in order to maintain a constant amount in the bank. Faced with a sudden bill and drop in the funds at the bank, he/she, when exercising proportional control, would deposit money in the bank in amounts proportional to the error, i.e. the amount short of the required amount.

The float method of controlling the level of water in a cistern is an example of the use of a proportional controller. The control mode is determined in this case by a lever. The error signal is the input to the ball end of the lever, the output is the movement of the other end of the lever (Figure 6.6). We have:

$$\text{output movement} = \frac{x}{y} \times \text{error}$$

The output is proportional to the error, the gain being x/y.

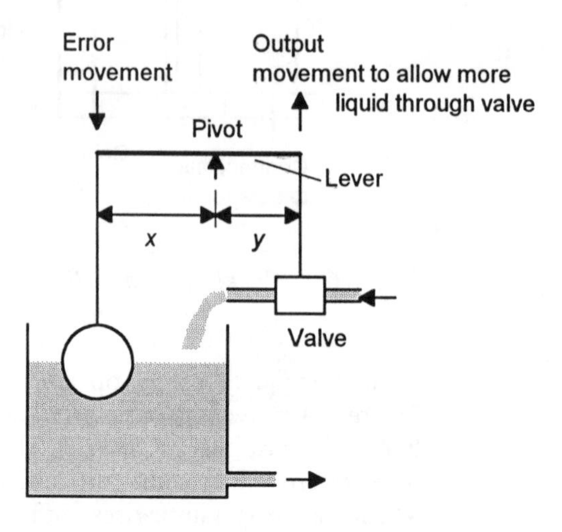

Figure 6.6 *The float-lever proportional controller*

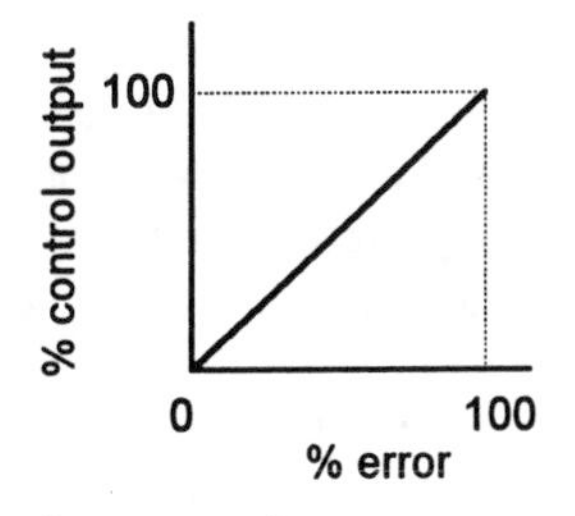

Figure 6.7 *Percentages*

Note that it is customary to express the output of a controller as a percentage of the full range of output that it is capable of passing on to the correction element. Thus, with a valve as a correction element, as in the float operated control of level in Figure 6.6, we might require it to be completely closed when the output from the controller is 0% and fully open when it is 100% (Figure 6.7). Because the controller output is proportional to the error, these percentages correspond to a zero value for the error and the maximum possible error value. When the error is 50% of its maximum value then the controller output will be 50% of its full range.

Another example of a proportional mode controller is an amplifier which gives an output which is proportional to the size of the input. Figure 6.8 illustrates, for the control of temperature of the outflow of liquid from a tank, the use of a differential amplifier as a comparison element and another amplifier as supplying the proportional control mode.

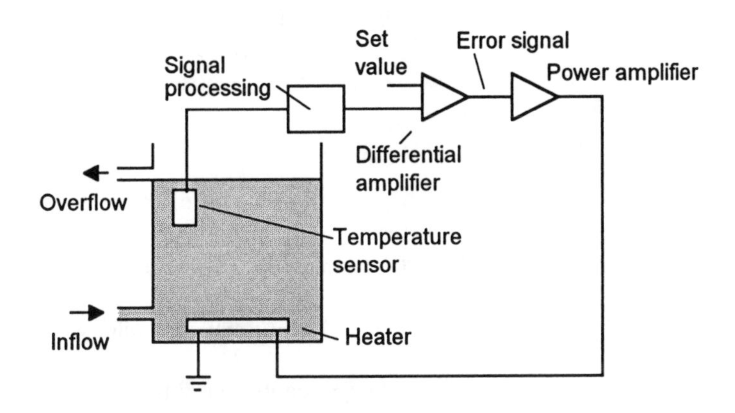

Figure 6.8 *Proportional controller for the control of temperature*

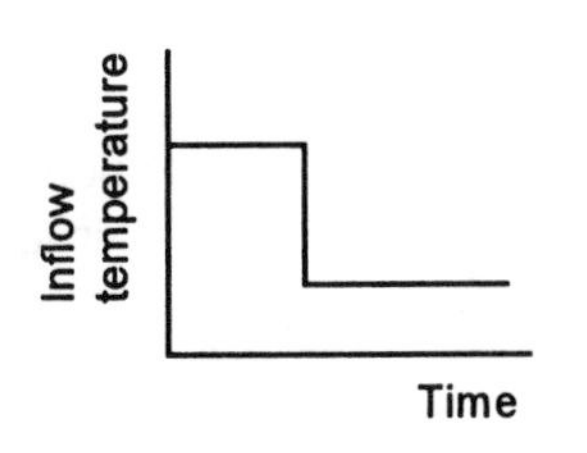

Figure 6.9 *Inflow change*

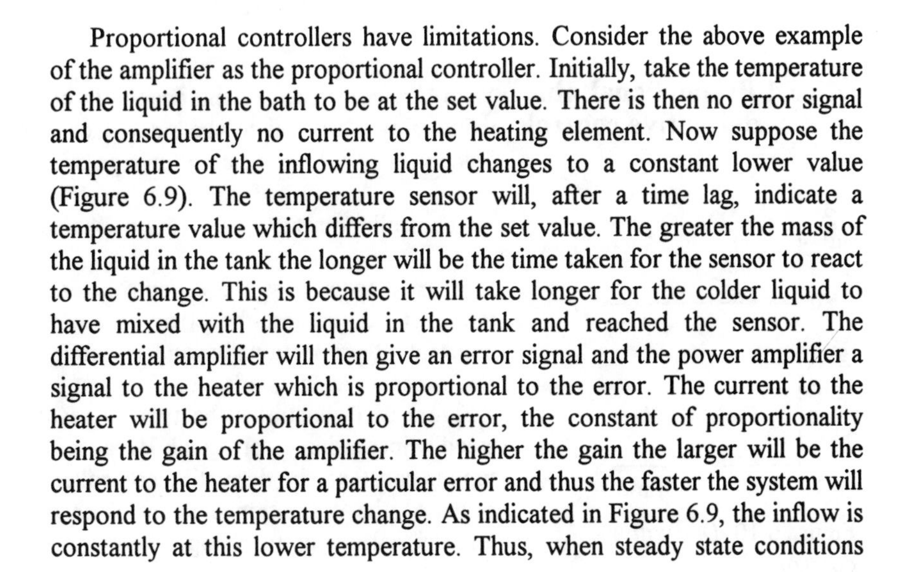

Proportional controllers have limitations. Consider the above example of the amplifier as the proportional controller. Initially, take the temperature of the liquid in the bath to be at the set value. There is then no error signal and consequently no current to the heating element. Now suppose the temperature of the inflowing liquid changes to a constant lower value (Figure 6.9). The temperature sensor will, after a time lag, indicate a temperature value which differs from the set value. The greater the mass of the liquid in the tank the longer will be the time taken for the sensor to react to the change. This is because it will take longer for the colder liquid to have mixed with the liquid in the tank and reached the sensor. The differential amplifier will then give an error signal and the power amplifier a signal to the heater which is proportional to the error. The current to the heater will be proportional to the error, the constant of proportionality being the gain of the amplifier. The higher the gain the larger will be the current to the heater for a particular error and thus the faster the system will respond to the temperature change. As indicated in Figure 6.9, the inflow is constantly at this lower temperature. Thus, when steady state conditions prevail, we always need current passing through the heater. Thus there must be a continuing error signal and so the temperature can never quite be the set value. This error signal which persists under steady state conditions is termed the *steady state error* or the *proportional offset*. The higher the gain of the amplifier the lower will be the steady state error because the system reacts more quickly.

In the above example, we could have obtained the same type of response if, instead of changing the temperature of the input liquid, we had made a sudden change of the set value to a new constant value. There would need to be a steady state error or proportional offset from the original value.

All proportional control systems have a steady state error. The proportional mode of control tends to be used in processes where the gain K_P can be made large enough to reduce the steady state error to an acceptable level. However, the larger the gain the greater the chance of the system oscillating and never settling down to a steady state value. The oscillations occur because of time lags in the system, the higher the gain the bigger will be the controlling action for a particular error and so the greater the chance that the system will overshoot the set value.

Example

A proportional controller has a gain of 4. What will be the percentage steady state error signal required to maintain an output from the controller of 20% when the normal set value is 0%?

With a proportional controller we can write:

$$\% \text{ controller output} = \text{gain} \times \% \text{ error}$$

$$20 = 4 \times \% \text{ error}$$

Hence the percentage error is 5%.

6.4 Proportional plus derivative control

With *derivative control* the change in controller output from the set point value is proportional to the rate of change with time of the error signal, i.e. controller output $\propto$ rate of change of error. Thus we can write:

$$\text{controller output} = K_D \times \text{rate of change of error}$$

It is usual to express these controller outputs as a percentage of the full range of output and the error as a percentage of full range. K_D is the constant of proportionality and is commonly referred to as the *derivative time* since it has units of time.

Error
0
Time
Controller output
0
Time

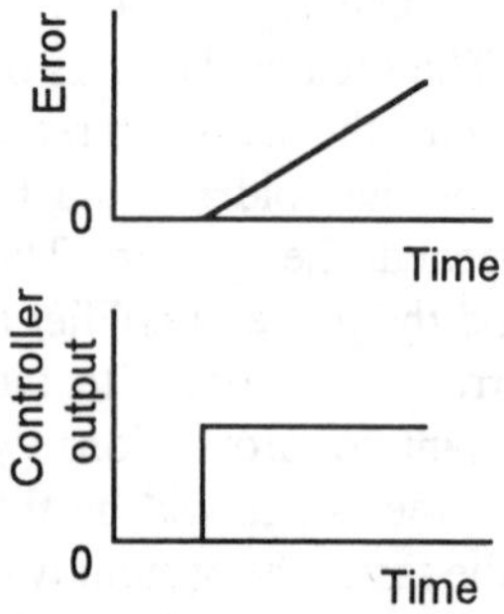

Figure 6.10 *Derivative control*

Figure 6.10 illustrates the type of response that occurs when there is a steadily increasing error signal. Because the rate of change of the error with time is constant, the derivative controller gives a constant controller output signal to the correction element. With derivative control, as soon as the error signal begins to change there can be quite a large controller output since it is proportional to the rate of change of the error signal and not its value. Thus with this form of control there can be rapid corrective responses to error signals that occur.

To illustrate the above, consider a person exercising control over his/her bank balance in order to maintain a constant amount in the bank. Faced with money constantly being withdrawn from the bank to meet bills, he/she, when exercising derivative control, would deposit money in the bank in amounts proportional to the rate at which the error, i.e. the amount short of the required amount, is changing.

Derivative controllers give responses to changing error signals but do not, however, respond to constant error signals, since with a constant error the rate of change of error with time is zero. Because of this, derivative control is combined with proportional control. Then we have:

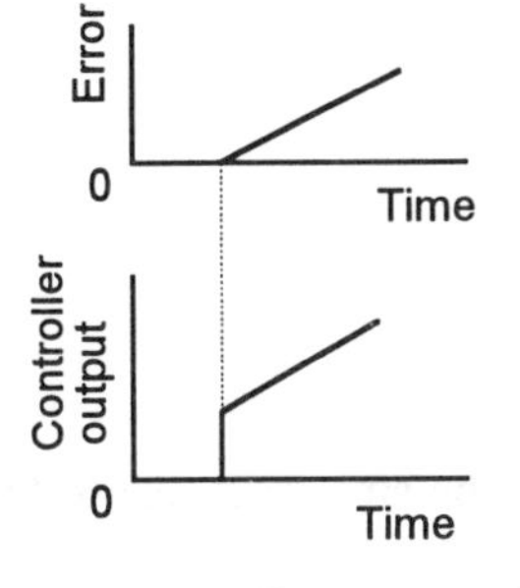

Figure 6.11 *Proportional plus derivative control*

$$\text{controller output} = K_P(\text{error} + K_D \times \text{rate of change of error with time})$$

Figure 6.11 shows how, with proportional plus derivative control, the controller output can vary when there is a constantly changing error. There is an initial quick change in controller output because of the derivative action followed by the gradual change due to proportional action. This form of control can thus deal with fast process changes better than just proportional control alone. It still, like proportional control alone, needs a steady state error in order to cope with a constant change in input conditions or a change in the set value.

Example

A derivative controller has a derivative constant K_D of 0.4 s. What will be the controller output when the error (a) changes at 2%/s, (b) is constant at 4%?

(a) Using the equation given above, i.e. controller output = K_D × rate of change of error, then we have:

$$\text{controller output} = 0.4 \times 2 = 0.8\ \%$$

This is a constant output.

(b) With a constant error there is no change of error with time and thus the controller output is zero.

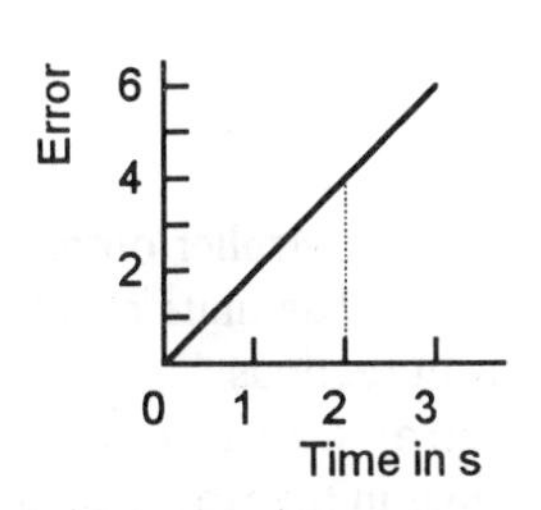

Figure 6.12 *Example*

Example

What will the controller output be for a proportional plus derivative controller (a) initially and (b) 2 s after the error begins to change from the zero error at the rate of 2%/s (Figure 6.12). The controller has $K_P = 4$ and $K_D = 0.4$ s.

(a) Initially the error is zero and so there is no controller output due to proportional action. There will, however, be an output due to derivative action since the error is changing at 2%/s. Since the output of the controller, even when giving a response due to derivative action alone, is multiplied by the proportional gain, we have:

$$\text{controller output} = K_P K_D \times \text{rate of change of error}$$

and so:

$$\text{controller output} = 4 \times 0.4 \times 2 = 3.2\%$$

(b) Because the rate of change is constant, after 2 s the error will have become 4%. Hence, then the controller output due to the proportional mode will be given by:

$$\text{controller output} = K_P \times \text{error}$$

and so that part of the output is:

$$\text{controller output} = 4 \times 4 = 16\%$$

The error is still changing and so there will still be an output due to the derivative mode. This will be given by

$$\text{controller output} = K_P K_D \times \text{rate of change of error}$$

and so:

$$\text{controller output} = 4 \times 0.4 \times 2 = 3.2\%$$

Hence the total controller output due to both modes is the sum of these two outputs and 16 + 3.2 = 19.2%.

6.5 Proportional plus integral control

Integral control is the control mode where the controller output is proportional to the integral of the error with respect to time, i.e.:

$$\text{controller output} \propto \text{integral of error with time}$$

and so we can write:

$$\text{controller output} = K_I \times \text{integral of error with time}$$

where K_I is the constant of proportionality and, when the controller output is expressed as a percentage and the error as a percentage, has units of s^{-1}. The reciprocal of K_I is called the *integral time* T_I and is in seconds.

To illustrate what is meant by the integral of the error with respect to time, consider a situation where the error varies with time in the way shown

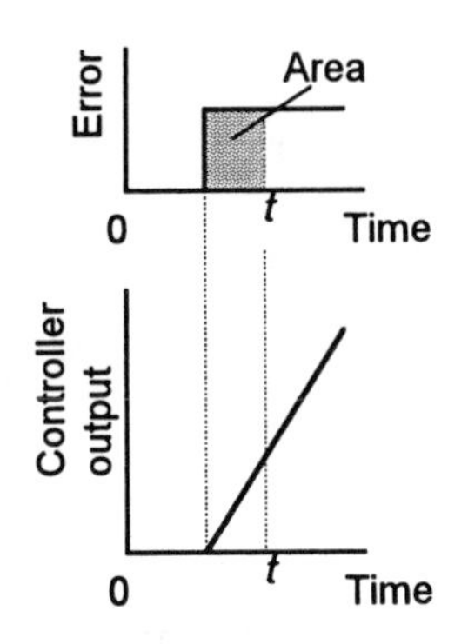

Figure 6.13 *Integral mode*

in Figure 6.13. The value of the integral at some time t is the area under the graph between $t = 0$ and t. Thus we have:

controller output $\propto$ area under the error graph between $t = 0$ and t

Thus as t increases, the area increases and so the controller output increases. Since, in this example, the area is proportional to t then the controller output is proportional to t and so increases at a constant rate. Note that this gives an alternative way of describing integral control as:

rate of change of controller output $\propto$ error

A constant error gives a constant rate of change of controller output.

To illustrate the above, consider a person faced with a sudden bill, i.e. an error in their funds. To meet this demand and exercise control over their money, when exercising integral control they would pay it off at a constant rate.

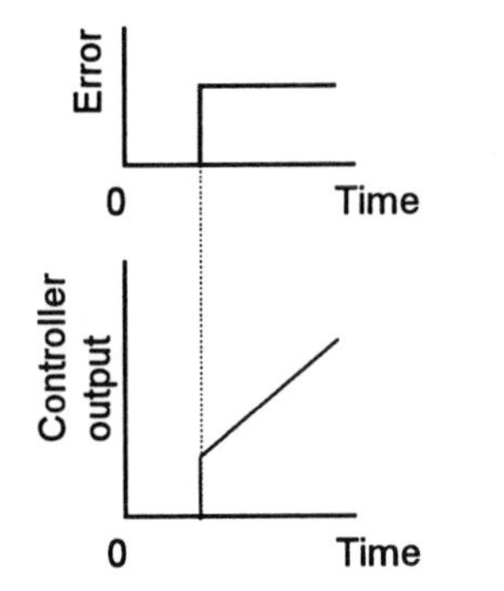

Figure 6.14 *Proportional plus integral mode*

The integral mode of control is not usually used alone but generally in conjunction with the proportional mode. When integral action is added to a proportional control system the controller output is given by:

controller output = K_P(error + integral of error with time)

where K_P is the proportional control constant and K_I the integral control constant. Figure 6.14 shows how the system reacts when there is an abrupt change to a constant error. The error gives rise to a proportional controller output which remains constant since the error does not change. There is then superimposed on this a steadily increasing controller output due to the integral action.

The combination of integral mode with proportional mode has one great advantage over the proportional mode alone: the steady state error can be eliminated. This is because the integral part of the control can provide a controller output even when the error is zero. The controller output is the sum of the area all the way back to time $t = 0$ and thus even when the error has become zero, the controller will give an output due to previous errors and can be used to maintain that condition. Figure 6.15 illustrates this.

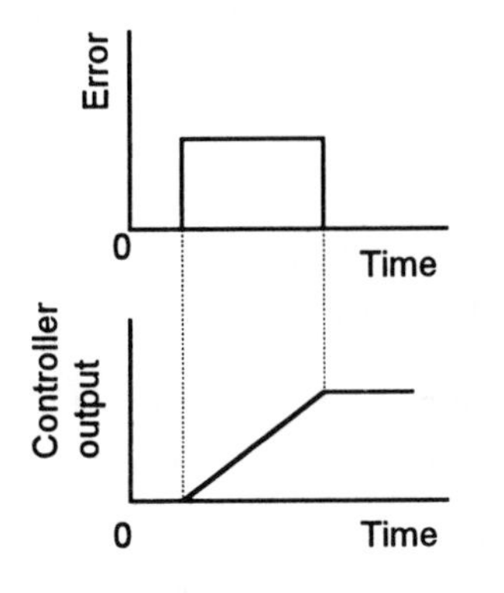

Figure 6.15 *Controller output when error becomes zero*

Because of the lack of a steady state error, this type of controller can be used where there are large changes in the process variable. However, because the integration part of the control takes time, the changes must be relatively slow to prevent oscillations.

Example

An integral controller has a value of K_I of 0.10 s^{-1}. What will be the output after times of (a) 1 s, (b) 2 s, if there is a sudden change to a constant error of 20%, as illustrated in Figure 6.16?

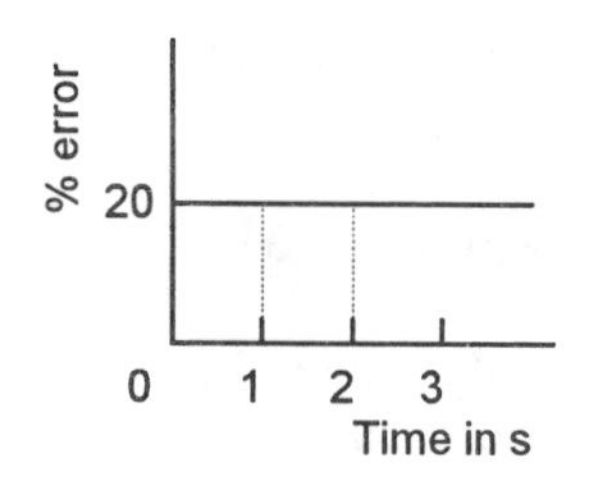

Figure 6.16 *Example*

We can use the equation:

controller output = K_I × integral of error with time

(a) The area under the graph between a time of 0 and 1 s is 20 %s. Thus the controller output is 0.10 × 20 = 2%.
(b) The area under the graph between a time of 0 and 2 s is 40 %s. Thus the controller output is 0.10 × 40 = 4%.

6.6 Three-mode control

Combining all three modes of control (proportional, integral and derivative) enables a controller to be produced which has no steady state error and reduces the tendency for oscillations. Such a controller is known as a *three-mode controller* or *PID controller*. The equation describing its action is:

controller output = K_P(error + K_I × integral of error + K_D × rate of change of error)

where K_p is the proportionality constant, K_I the integral constant and K_D the derivative constant. A three-mode controller can be considered to be a proportional controller which has integral control to eliminate the offset error and derivative control to reduce time lags.

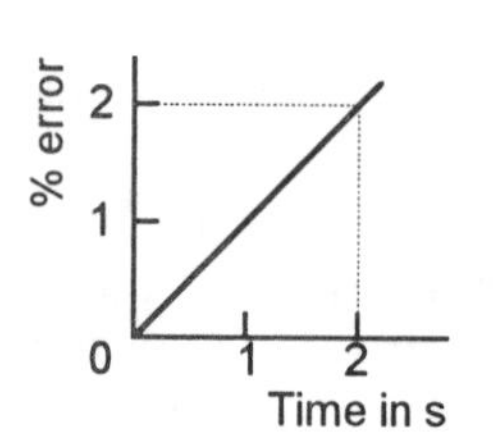

Figure 6.17 *Example*

Example

Determine the controller output of a three-mode controller having K_P as 4, K_I as 0.6 s^{-1}, K_D as 0.5 s at time (a) t = 0 and (b) t = 2 s when there is an error input which starts at 0 at time t = 0 and increases at 1%/s (Figure 6.17).

(a) Using the equation:

controller output = K_P(error + K_I × integral of error + K_D × rate of change of error)

we have for time t = 0 an error of 0, a rate of change of error with time of 1 s^{-1}, and an area between this value of t and t = 0 of 0. Thus:

controller output = 4(0 + 0 + 0.5 × 1) = 2.0%

(b) When t = 2 s, the error has become 1%, the rate of change of the error with time is 1%/s and the area under between t = 2 and t = 0 is 1%s. Thus:

controller output = 4(1 + 0.5 × 1 + 1) = 10%

6.7 System response and tuning

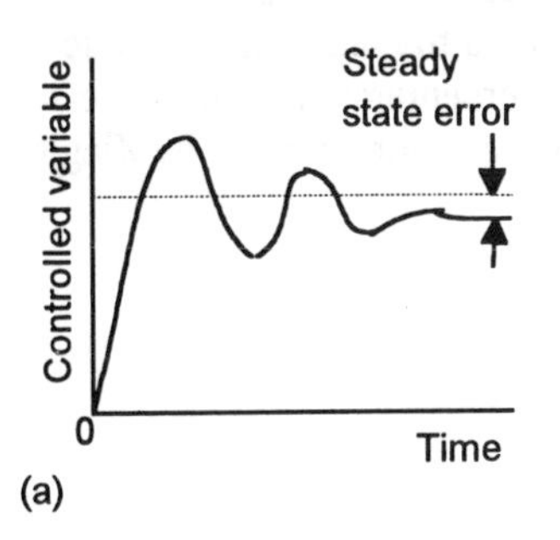

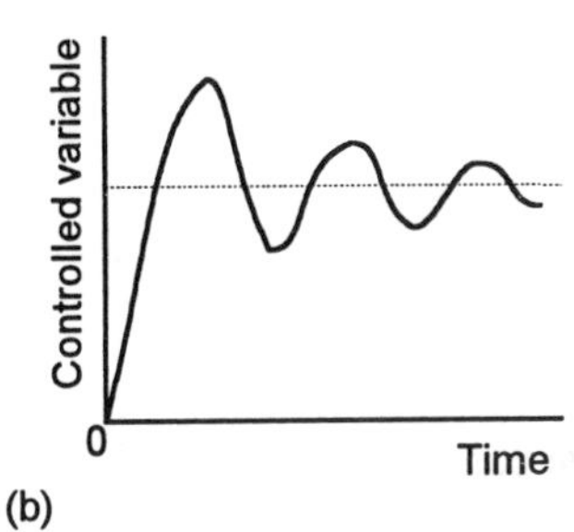

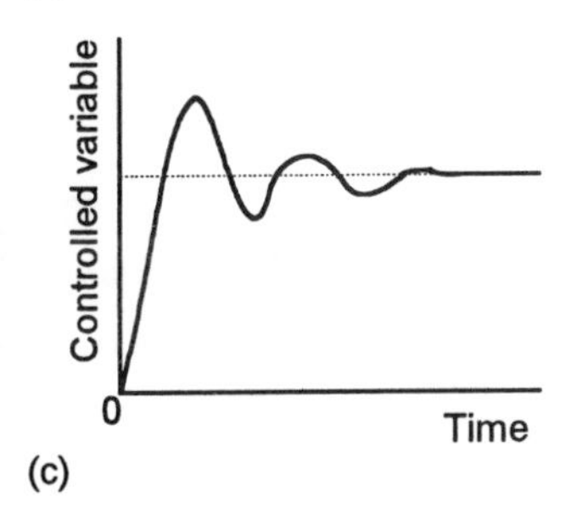

Figure 6.18 *Responses to (a) P, (b) PI, (c) PID control.*

The design of a controller for a particular situation involves selecting the control modes to be used and the control mode settings. This means determining whether proportional control, proportional plus derivative, proportional plus integral or proportional plus integral plus derivative is to be used and selecting the values of K_P, K_I and K_D. These determine how the system reacts to a disturbance or a change in the set value, how fast it will respond to changes, how long it will take to settle down after a disturbance or change to the set value, and whether there will be a steady state error.

Figure 6.18 illustrates the types of response that can occur with the different modes of control when subject to a step input, i.e. a sudden change to a different constant set value or perhaps a sudden constant disturbance. Proportional control gives a fast response with oscillations which die away to leave a steady state error. Proportional plus integral control has no steady state error but is likely to show more oscillations before settling down. Proportional but integral plus derivative control has also no steady state error, because of the integral element, and is likely to show less oscillations than the proportional plus integral control. The inclusion of derivative control reduces the oscillations.

The term *tuning* is used to describe methods used to select the best controller setting by means of testing the actual control system. The following are two methods that can be used, both by *Ziegler and Nichols*.

6.7.1 Process reaction method

This method uses certain measurements made from testing the system with the control loop open so that no control action occurs. Generally the break is made between the controller and the correction unit (Figure 6.19). A test input signal is then applied to the correction unit and the response of the controlled variable determined.

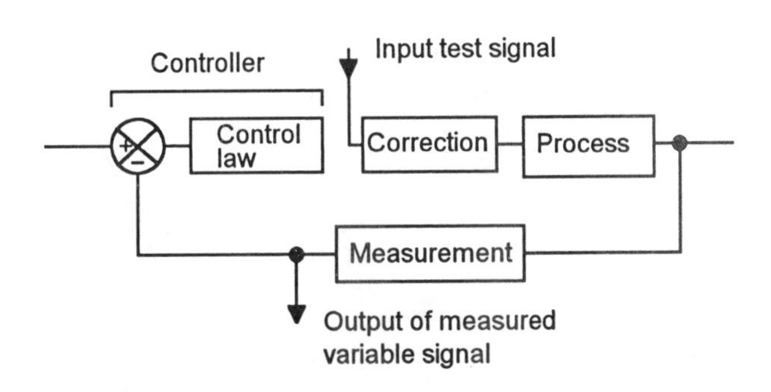

Figure 6.19 *Test arrangement*

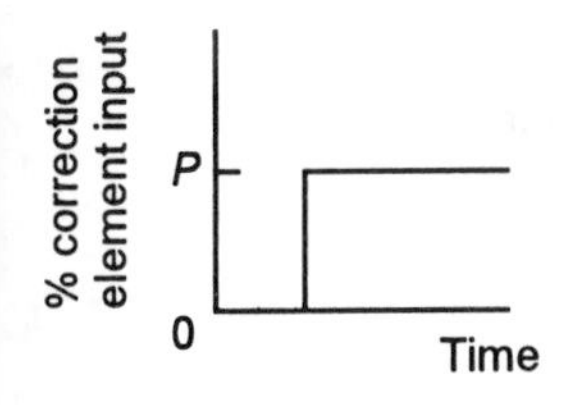

Figure 6.20 *Test signal*

The test signal is a step signal with a step size expressed as the percentage change P in the correction unit (Figure 6.20). The response of the controlled variable to such an input is monitored and a graph (Figure

6.21) of the variable plotted against time. This graph is called the *process reaction curve*. A tangent is drawn to give the maximum gradient of the graph. The time between the start of the test signal and the point at which this tangent intersects the graph time axis is termed the lag L. If the value of the maximum gradient is M, expressed as the percentage change of the set value of the variable per minute, then the criteria recommended by Ziegler and Nichols for control settings are:

1 For proportional control

$$K_P = \frac{P}{ML}$$

2 For proportional plus integral control

$$K_P = 0.9\left(\frac{P}{ML}\right)$$

$$K_I = \frac{0.3}{L} \text{ min}^{-1}$$

3 For proportional plus integral plus derivative control

$$K_P = 1.2\left(\frac{P}{ML}\right)$$

$$K_I = \frac{0.5}{L} \text{ min}^{-1}$$

$$K_D = 0.5L \text{ min}$$

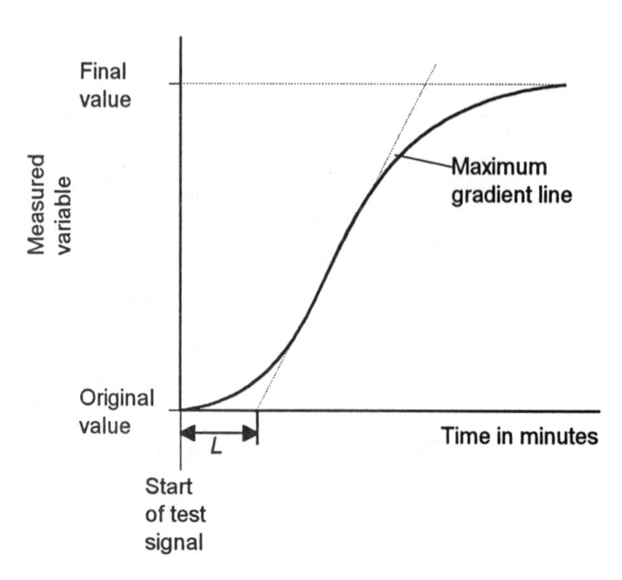

Figure 6.21 *The process reaction graph*

Example

Determine the settings of K_P, K_I and K_D required for a three-mode controller which gave a process reaction curve shown in Figure 6.22 when the test signal was a 10% change in the control valve position.

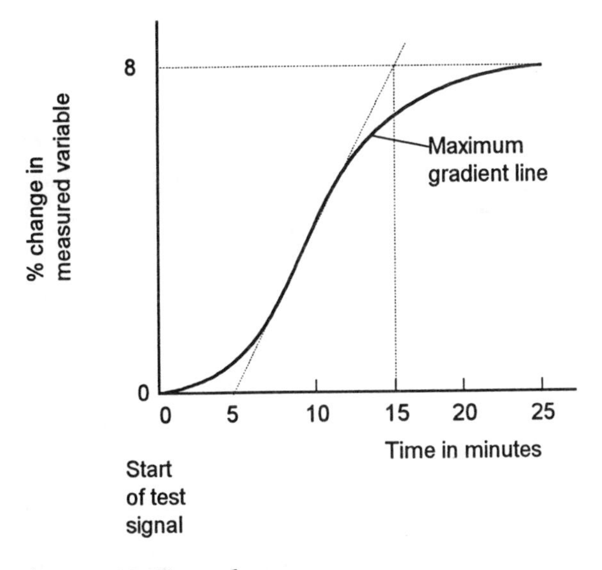

Figure 6.22 *Example*

Drawing a tangent to the maximum gradient part of the graph gives a lag L of 5 minutes and a gradient M of 8/10 = 0.8 %/min. Hence

$$K_P = \frac{1.2P}{ML} = \frac{1.2 \times 10}{0.8 \times 5} = 3$$

$$K_I = \frac{0.5}{L} = \frac{0.5}{5} = 0.1 \text{ min}^{-1}$$

$$K_D = 0.5L = 0.5 \times 5 = 2.5 \text{ min}$$

6.7.2 Ultimate cycle method

With this method, the integral and derivative actions are first reduced to their least effective values. The proportional constant K_P is then set low and gradually increased until oscillations in the controlled variable start to occur. The critical value of the proportional constant K_{Pc} at which this occurs is noted and the periodic time of the oscillations T_c measured. The Ziegler and Nichols recommended criteria for controller settings are then:

1 For proportional control

$$K_P = 0.5K_{Pc}$$

2 For proportional plus integral control

$$K_P = 0.45K_{Pc}$$

$$K_I = \frac{1.2}{T_c}$$

3 For proportional plus integral plus derivative control

$$K_P = 0.6K_{Pc}$$

$$K_I = \frac{2.0}{T_c}$$

$$K_D = \frac{T_c}{8}$$

Example

When tuning a three-mode control system by the ultimate cycle method it was found that, with derivative and integral control switched off, oscillations begin when the proportional gain is increased to 3.3. The oscillations have a periodic time of 500 s. What are the suitable values of K_P, K_I and K_D?

Using the equations given above:

$$K_P = 0.6K_{Pc} = 0.6 \times 3.3 = 1.98$$

$$K_I = \frac{2}{T_c} = \frac{2}{500} = 0.004\ \text{s}^{-1}$$

$$K_D = \frac{T_c}{8} = \frac{500}{8} = 62.5\ \text{s}$$

6.8 Process controllers

Controller hardware for use in analogue control loops for process control is normally calibrated to give displays with scales reading 0 to 100% of the signal variation. On one scale can be displayed the set point and the actual instantaneous value of the controlled variable, while on another scale is diplayed the controller output signal. Controls are also supplied for adjusting the proportional, integral and derivative constants. Figure 6.23 illustrates the type of displays commonly obtained and the way the controller is connected into the control loop.

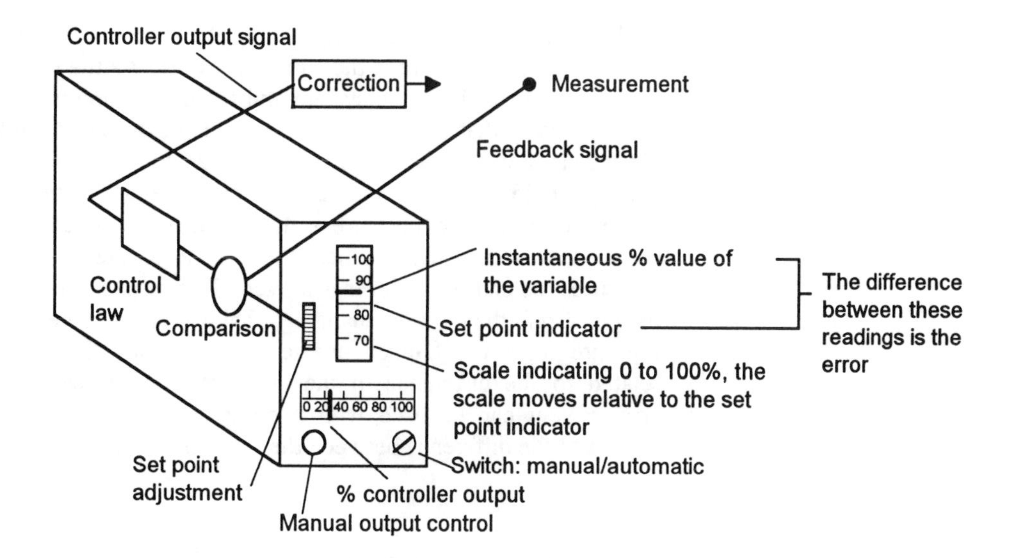

Figure 6.23 *Typical form of an analogue controller*

A switch is generally provided to allow for the controller to be switched from automatic to manual control and allow the controller output to be set directly by the operator. The closed-loop is then broken and the system is open-loop. Such control is useful during the tuning of the system (see section 6.7.1), and during the start-up and shut-down of plant.

6.8.1 Pneumatic controller

The basic elements in pneumatic controllers are bellows and a flapper-nozzle arrangement. When the air pressure in bellows is changed the length changes and this can be used to rotate levers (Figure 6.24(a)). The flapper-nozzle arrangement (Figure 6.24(b)) gives an output pressure that is related

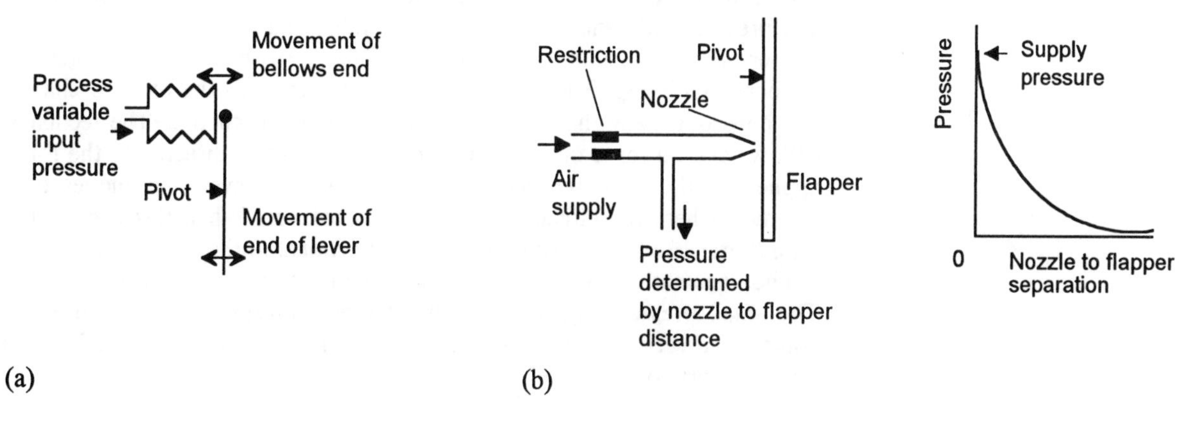

Figure 6.24 *(a) Bellows, (b) flapper–nozzle*

to the distance between the flapper and the nozzle. When the separation is zero the output pressure is the supply pressure. As the separation increases so more air leaks from the system and the output pressure drops. The type of relationship that occurs between output pressure and the flapper–nozzle distance is shown in the graph in Figure 6.24(b)).

Figure 6.25 shows how a pair of bellows and a flapper–nozzle can be combined to give a pressure output which is related to the difference between two pressures. When there is a difference between the two pressures the end of the lever is moved. This results in the separation of the flapper from the nozzle changing. This, in turn, determines the output pressure from the flapper–nozzle system. Thus the output pressure is related to the difference in the pressures in the two bellows. Such an arrangement can be used with a control system to give an error signal related to the difference between the set pressure and the actual pressure.

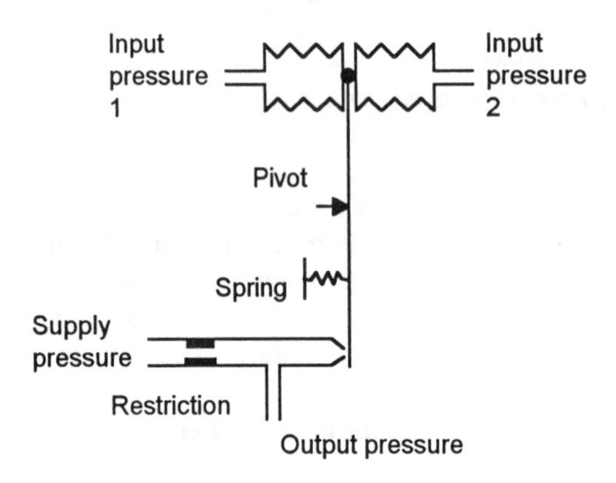

Figure 6.25 *Differential pressure system*

Figure 6.26 shows the basic form of a pneumatic proportional controller. To illustrate the action of the controller, consider what happens when there is a pressure difference between the set value and the measured pressure. The error signal results in a movement of the lever which then changes the flapper–nozzle gap. This then causes the air supply to the relay to change and consequentially that to the feedback bellows to change. The feedback bellows exert a force on the lever which is oppositely directed to the force applied by the error sensing pair of bellows. The lever thus moves to a situation where the effects of these forces balance. When this occurs, the output pressure from the controller is proportional to the error signal.

The pneumatic relay is a device similar in action to the electrical relay. The small pressure signals from the flapper–nozzle system are used to control and vary the much larger pressure of the supply. Figure 6.27 shows the basic form of such a relay.

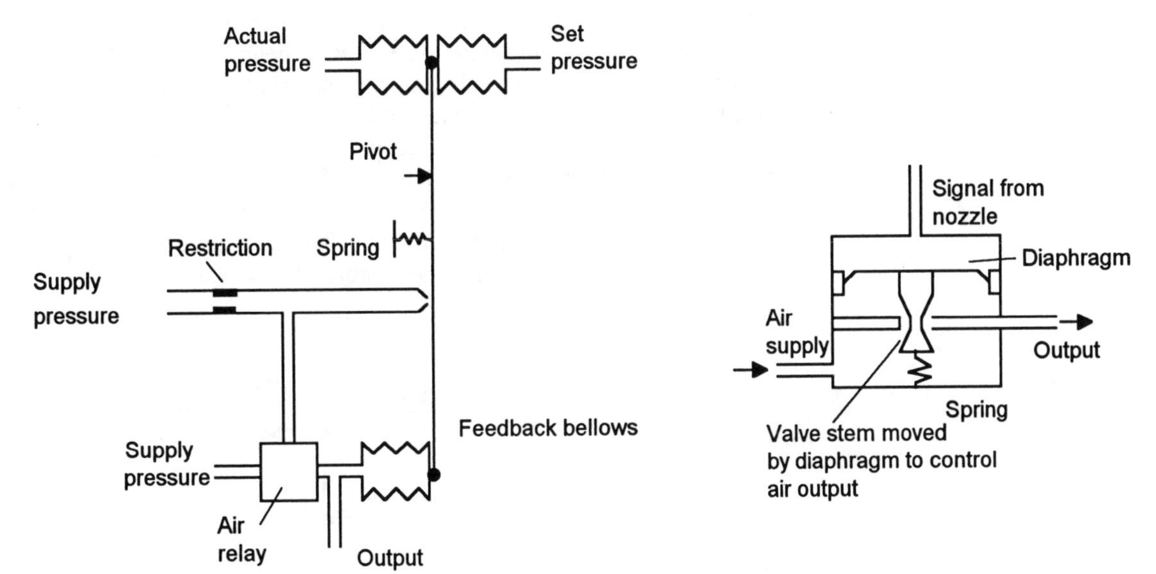

Figure 6.26 *Proportional controller*

Figure 6.27 *Pneumatic relay*

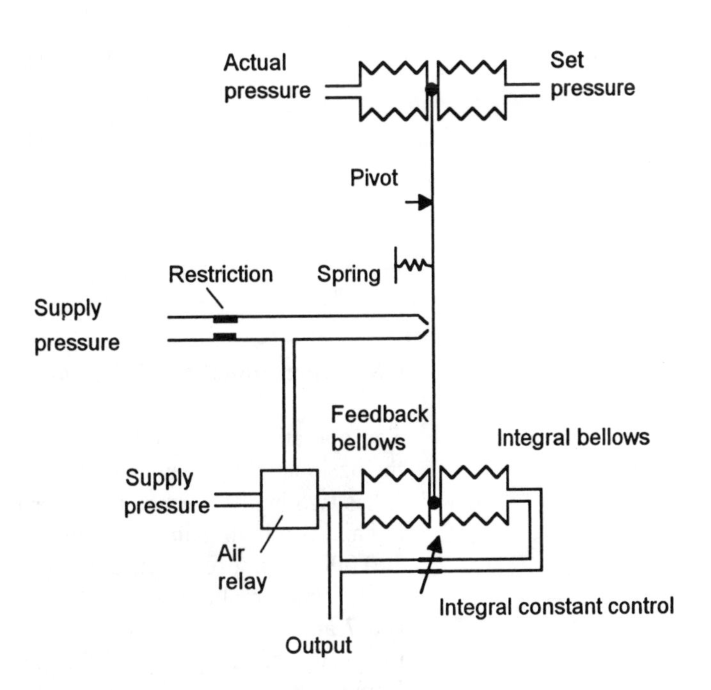

Figure 6.28 *Proportional plus integral controller*

Proportional plus integral control is obtained by the use of a controller of the form shown in Figure 6.28. The integral bellows oppose the action of the feedback bellows, responding to the changes in pressure from the relay via a restriction. The restriction introduces a time factor into the response of the integral bellows. The result is a controller output which is proportional to the error and the integral of the error with respect to time.

Figure 6.29 shows how a three mode controller is produced, i.e. proportional plus integral plus derivative. The derivative action is produced by incorporating a restriction in the feedback bellows line.

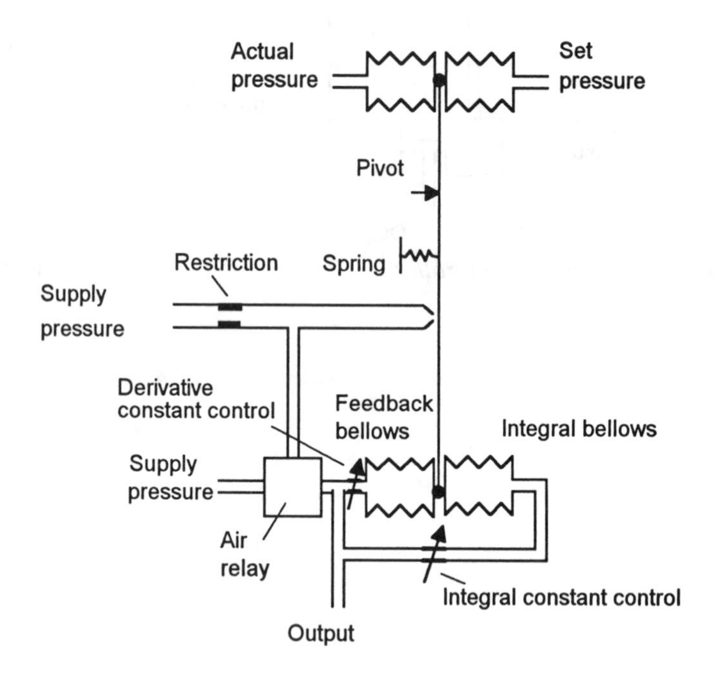

Figure 6.29 *PID controller*

6.8.2 Operational amplifier controller

The operational amplifier is the basis of many signal processing elements, the basic amplifier being supplied as an integrated circuit on a silicon chip. It has two inputs, termed the inverting input (–) and the non-inverting input (+) and is a high gain d.c. amplifier, the gain typically being of the order of 100 000 or more. Figure 6.30 shows the pin connections for a 741 operational amplifier with the symbol for the operational amplifier. Pins 4 and 7 are for the connections to the supply voltage for the amplifier, pin 2 for the inverting input, pin 3 for the non-inverting input. The output is taken from pin 6. Pins 1 and 5 are for the offset null. These are to enable circuits to be connected to enable corrections to be made for the non-ideal behaviour of the amplifier.

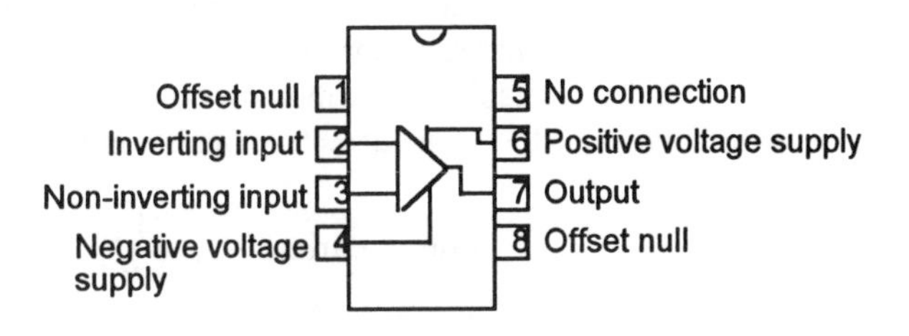

Figure 6.30 *The 741 operational amplifier*

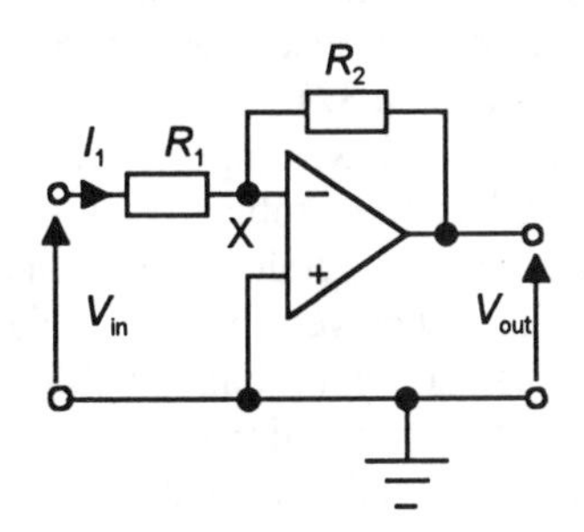

Figure 6.31 *Inverting amplifier*

Consider the amplifier when used as an *inverting amplifier* (Figure 6.31), i.e. an amplifier which gives an output which is out-of-phase with respect to the input. For the circuit shown in Figure 6.31, the connections for the power supply and the offset null have been omitted. The input is connected to the inverting input, the non-inverting input being connected to earth. A feedback loop is connected, via the resistor R_2, to the inverting input. The output voltage of such an amplifier is limited to about ±10 V and thus, since the gain is about 100 000, the input voltage to the inverting input at X, V_X, must be between about +0.0001 V and −0.0001 V. This is virtually zero and so point X is at virtually earth potential. For this reason it is called a *virtual earth*. The potential difference across the input resistance R_1 is $(V_{in} - V_X)$ and thus $(V_{in} - V_X) = I_1R_1$. But V_X is virtually zero and so we can write:

$$V_{in} = I_1R_1$$

Operational amplifiers have very high resistance between their input terminals, e.g. the resistance with the 741 operational amplifier is about 2 MΩ. Thus virtually no current flows from point X through the inverting input and so to earth. Thus the current I_1 that flows through R_1 must be essentially the current flowing through R_2. The potential difference across R_2 is $(V_X - V_{out})$. Thus we can write $(V_X - V_{out}) = I_1R_2$. But as V_X is effectively zero, we can write:

$$-V_{out} = I_2R_2$$

Eliminating I_1 from these two simultaneous equations gives:

$$\text{gain of circuit} = \frac{V_{out}}{V_{in}} = -\frac{R_2}{R_1}$$

The negative sign indicates that the output is 180° out-of-phase with the input. The gain is determined solely by the values of the two resistors.

A non-inverting amplifier can likewise be produced by taking the input to the non-inverting input instead of the inverting input.

As an example of the use of the above circuit for signal processing, consider a *voltage-to-current converter*. Situations can arise where the output from a stage may need to be a current in order to perhaps drive an electromechanical device such as a relay, stepper motor or perhaps give a display on a moving coil meter. If we use the coil of such a device as R_2 then the current through it is I_1 and so, since $V_{in} = I_1R_1$, then:

$$\text{output current} = I_1 = \frac{V_{in}}{R_1}$$

The input voltage V_{in} has been converted to the output current I_1. This circuit represents a very basic form of the circuits used for voltage-to-current converters.

We can also have the requirement for signal processing to convert a current to a voltage. Essentially a *current-to-voltage converter* consists of a resistor through which the current is passed (Figure 6.32). The potential difference across the resistor is then, assuming Ohm's law, proportional to the current. However, this simple arrangement has the problem that the load across which this voltage is applied is in parallel with the voltage-to-current converter resistance and if this is not very high it could shunt the resistance. To overcome this we can use an operational amplifier circuit. We then have the high input resistance of the operational amplifier between the voltage-to-current converter resistance and the load. Figure 6.33 shows the basic form such a circuit might take. The current is, as before, used to produce a potential difference across a resistance. The amplifier then transforms this voltage to a voltage output across a high resistance.

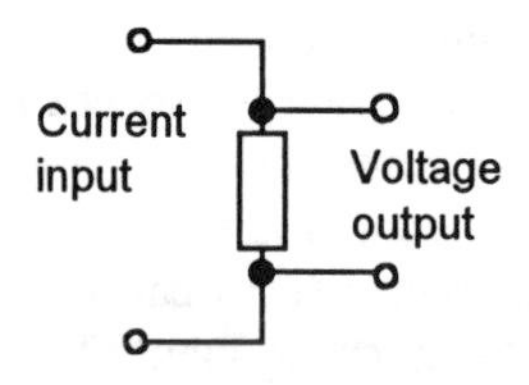

Figure 6.32 *Current-to-voltage converter*

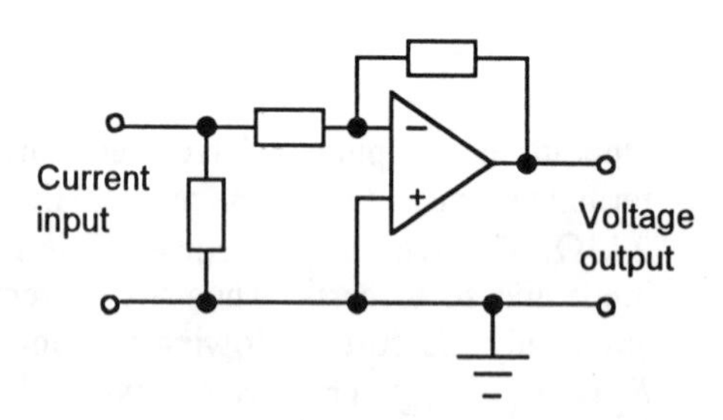

Figure 6.33 *A current-to-voltage converter*

Consider now how an *electronic controller* can be produced. The first step is to use a circuit to give the error signal, i.e. the difference between the set value and the actual value. Figure 6.34 shows how this can be done with a circuit involving just resistors. The value of the set value voltage is set by moving the slider on the variable resistor. This determines the fraction of the fixed voltage V which appears between the slider and earth. Added to this, and in the opposite direction, is the actual voltage that is fed back from the measurement part of the control system. The result is that the output is the difference between the set value and the actual value, i.e. the error voltage.

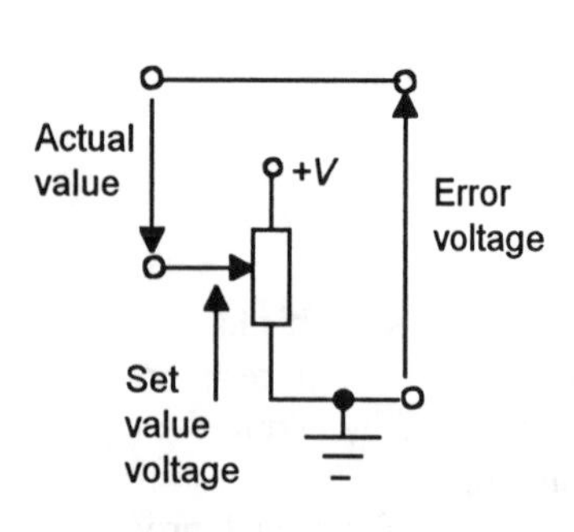

Figure 6.34 *Error detector*

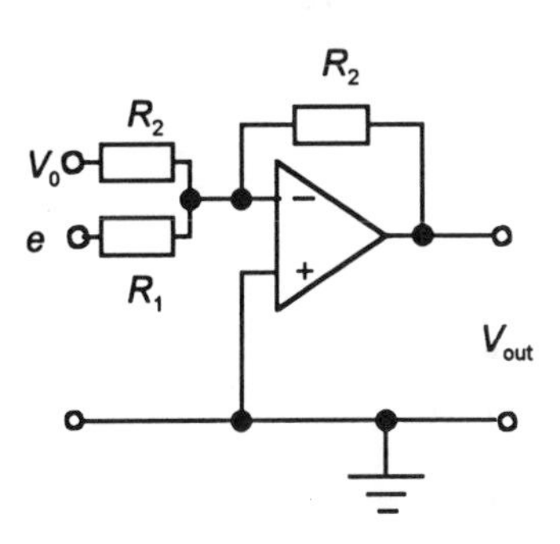

Figure 6.35 *Proportional controller with offset*

Having obtained the error signal, the next stage in an electronic controller is to implement the control law. Since the proportional mode requires just amplification of the error signal, an operational amplifier connected as an inverting amplifier can be used. If the proportional controller requires an output when there is zero error, e.g. in control of the rate of flow of a liquid there might be a requirement for the valve controlling the rate of flow to be 50% open with zero error, then the operational amplifier circuit can be modified to the form shown in Figure 6.35. The circuit has two inputs, the error voltage e and the value of the voltage V_0 required to give the output which is to occur with zero error, the so-termed offset voltage. The current given by the potential difference (error voltage e – virtual earth voltage) across R_1 is effectively e/R_1. The current given by the potential difference (offset voltage V_0 – virtual earth voltage) across the input resistance R_2 is effectively V_0/R_2. Since the operational amplifier has a very high resistance, the sum of these currents will be effectively the current through the feedback resistance R_2. Thus:

$$\text{current through } R_2 = \frac{e}{R_1} + \frac{V_0}{R_2}$$

The voltage across the feedback resistor, and hence the output voltage, is thus:

$$\text{output voltage } V_{\text{out}} = R_2\left(\frac{e}{R_1} + \frac{V_0}{R_2}\right) = \left(\frac{R_2}{R_1}\right)e + V_0$$

Thus the output is proportional to the error but offset by the voltage V_0.

Figure 6.36 shows how an operational amplifier can be used to give the proportional plus integral mode. The feedback path now has a capacitor as well as a resistor. With no error signal there will be no input voltage to the operational amplifier and so no current through the feedback path to charge the capacitor. Suppose now we have an error which suddenly increases to some value. The result will be a current through R_1 and hence the feedback path. Current is the rate of movement of charge and so the result is that the capacitor becomes charged, the rate at which it becomes charged being determined by the resistance in the circuit and the value of the capacitance. The proportional gain is R_2/R_1 and the integral constant $1/R_2C$.

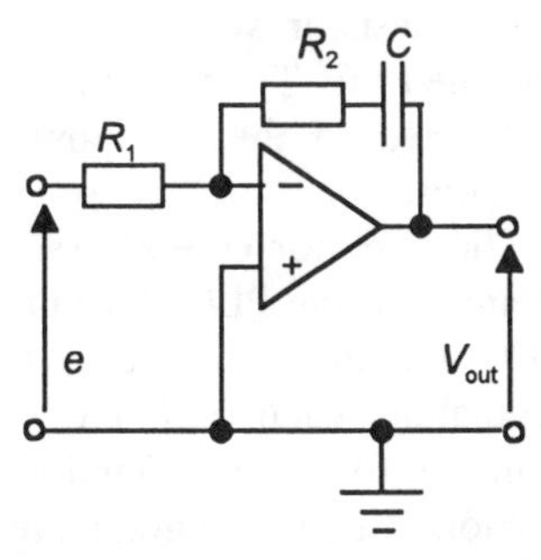

Figure 6.36 *Proportional plus integral mode*

Figure 6.37 shows how an operational amplifier can be used to give the proportional plus integral plus derivative mode. The proportional gain is $R_2/(R_1 + R_D)$, the integral constant $1/R_2C_I$ and the derivative constant R_DC_D.

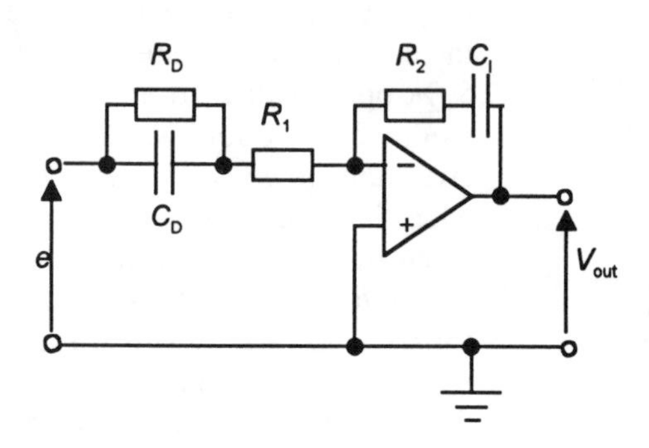

Figure 6.37 *Proportional plus integral plus derivative controller*

6.9 Programmable logic controllers

A *programmable logic controller* (PLC) (see section 4.3) is an electronic device that uses a programmable memory to store instructions and to implement functions such as logic, sequencing, timing, counting and arithmetic in order to control machines and processes. The term *logic* is used because programming is primarily concerned with implementing logic operations such as OR, AND, etc. in switching circuits.

The input devices such as switches and sensors which respond to the conditions occurring and the output devices in the system being controlled, e.g. motors, are connected to the PLC. The input/output channels of the PLC provide signal processing and isolation functions so that sensors and actuators can be generally directly connected to them without the need for other circuitry. Outputs are often specified as being of relay type, transistor type or triac type. With the relay type, the signal from the PLC output is used to operate a relay and so is able to switch currents of the order of a few amperes in an external circuit. The relay isolates the PLC from the external circuit but is relatively slow to operate. The transistor type of output uses a transistor to switch current through the external circuit. This gives a faster switching action. Optoisolators (see Figure 2.53) are used with transistor switches to provide isolation between the external circuit and the PLC. Triac outputs can be used to control external loads which are connected to the a.c. power supply with optoisolators used to provide isolation.

The operator enters a sequence of instructions, i.e. a program, into the memory of the PLC. This program determines the control behaviour of the PLC. Its aims is to specify the precise conditions for turning on or off, or regulating, each output of the controller. The program can be specified using a form of programming termed *ladder programming* (see next section). Such a program might be entered into the computer memory using a programming terminal which is plugged into the PLC (Figure 6.38). The

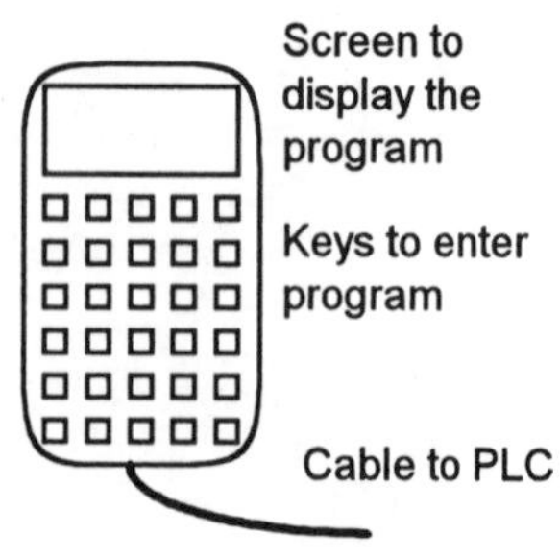

Figure 6.38 *A programming terminal*

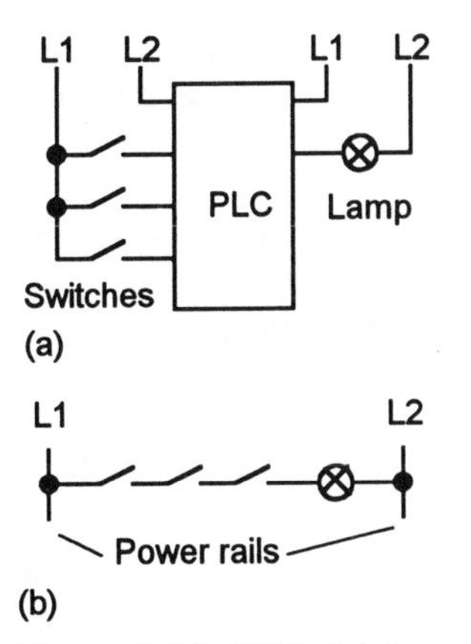

Figure 6.39 *PLC: (a) inputs and output, (b) effective circuit*

PLC then monitors the inputs and outputs according to this program and controls the machine or process concerned.

To illustrate the above, consider a simple example. A lamp is to come on when three switches are closed. Figure 6.39(a) shows the inputs, the switches, and the output, the lamp, connected to the PLC. L1 and L2 are the connections to the power supply. The program to be entered has then to instruct the PLC to effectively 'connect' the three inputs in series with the lamp so that power is applied to the lamp when all three switches are closed. Figure 6.39(b) shows the effective circuit with the switches and lamp connected in series between the power lines. This might then be one event in a controlled sequence.

The programmable logic controller is designed to carry out the logic functions previously carried out by such components as relays, mechanical timers such as cams, etc. PLCs have the great advantage that it is possible to modify a control system without having to rewire the input and output devices, the only requirement being that an operator has to key in a different set of instructions. The result is a flexible system which can be used to control systems which vary quite widely in their nature and complexity.

6.9.1 Ladder programs

A ladder diagram is a simple way of specifying the controlled sequence in event driven processes. Such a diagram consists of two vertical lines and any number of horizontal lines connecting the vertical lines. It looks like a ladder. Each rung on the ladder defines one step in the control process. Writing a program is equivalent to drawing a sequence of switching circuits. The two vertical lines of the ladder represent the power rails with circuits connected as the horizontal lines, i.e. the rungs of the ladder, between these two verticals. Figure 6.39(b) illustrates this, giving one rung of such a ladder.

In drawing a ladder program certain conventions are adopted:

1 A ladder diagram is read from left to right and from top to bottom.
2 The vertical lines of the ladder diagram represent the power rails.
3 Each rung on the ladder defines one operation in the process.
4 Inputs must always precede outputs and there must be at least one output on each line.
5 Each rung must start with an input or a inputs and end with an output.
6 Electrical devices are shown in their normal condition. This normal condition is the state they would be in when the electricity if off and no external forces are acting on them. Thus a switch which is open when there is no current or no object forcing it to be closed would be shown as open.
7 A particular device can appear in more than one rung of a ladder, e.g. a relay which switches on more than one device, and is designated by the same letters or numbers on each occasion.
8 Figure 6.40 shows some of the standard symbols used in ladder diagrams.

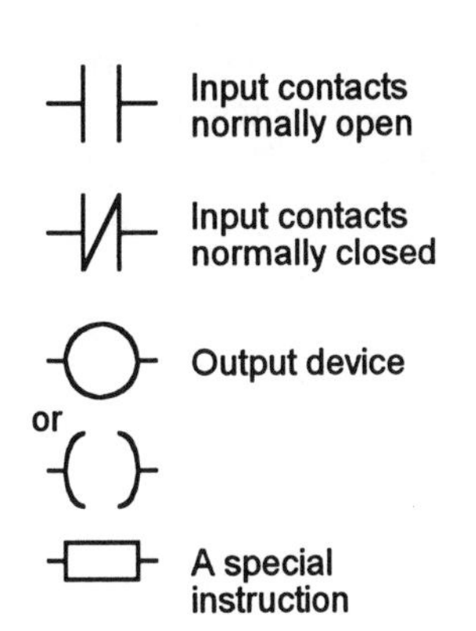

Figure 6.40 *Standard symbols*

The sequence followed by a PLC when carrying out a ladder program involves each rung of the ladder being scanned in turn. The sequence followed is thus:

1 Scan the inputs associated with the first rung, the top one, of the ladder program.
2 Solve the logic operation involving the inputs for that rung.
3 Set/reset the outputs for that rung.
4 Move on to the next rung and repeat operations 1, 2, 3.
5 Move on to the next rung and repeat operations 1, 2, 3.

⋮

And so on, until the end of the program, i.e. the end, bottom, rung on the ladder, is reached.

The inputs and outputs are numbered, the notation used depending on the PLC manufacturer. The Mitsubishi PLCs use the notation of writing an X before input elements and a Y before output elements. The following illustrate the types of numbers used to specify individual inputs and outputs with a typical Mitsubishi PLC:

Inputs X400–407, 410–413
X500–507, 510–513
(A total of 24 possible inputs)

Outputs Y430–437
Y530–537
(A total of 16 possible outputs)

Figure 6.41 *An example of a ladder rung*

To illustrate the drawing of a ladder diagram, consider a situation where the starting of a motor output device depends on a normally open start switch being closed. Starting with the input, we represent the normally open switch by the symbol | |. This might be labelled X400. The ladder rung terminates with the output, the motor, which is designated by the symbol O. This might be labelled Y430. We thus have the ladder diagram shown in Figure 6.41. When the switch is closed the motor is activated.

As another example, consider a situation where a motor is started when two, normally open, switches have both to be closed. This might represent a machine tool motor which will not start until the power switch is on and the switch indicating the closure of the safety guard is on. This describes an AND logic gate situation. The truth table is (see section 5.4):

Inputs		Output
Switch A	Switch B	
0	0	0
0	1	0
1	0	0
1	1	1

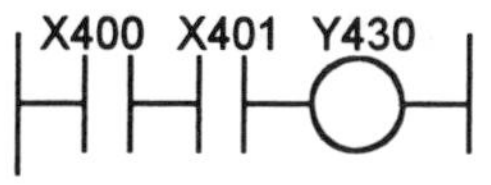

X400 X401 Y430

Figure 6.42 *AND logic gate*

The ladder diagram starts with | |, labelled X400, to represent switch A and in series with it | |, labelled X401, to represent switch B. The line then terminates with O, labelled Y430, to represent the output. Figure 6.42 shows the ladder rung.

Figure 6.43 shows a situation where a motor is not switched on until either, normally open, switch A or switch B is closed. The situation is an OR logic gate. The truth table for this arrangement is:

Inputs		Outputs
Switch A	Switch B	
0	0	0
0	1	1
1	0	1
1	1	1

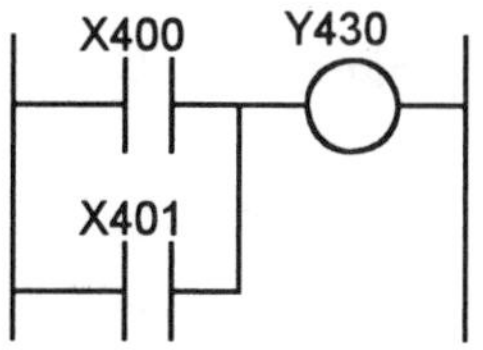

X400 Y430 X401

Figure 6.43 *OR logic gate*

The ladder diagram starts with | |, labelled X400, to represent switch A and in parallel with it | |, labelled X401, to represent switch B. The line then terminates with O, labelled Y430, to represent the output.

Consider another example of a program required to start a device when switch A and switch B are closed and keep the device operating when switch A is no longer closed. The device is to stop when B is opened. The two switches might be push buttons with A pushed to start and B pushed to stop. Figure 6.44 shows the program rung. Switch A is represented by X400 and is normally open. Switch B is represented by X401 and is normally closed. When switch A is closed then the device, Y430, is activated. But Y430 has a set of contacts associated with it and which close when it is activated. We thus have an OR gate between X400 and the Y430 contacts. Such a type of circuit is known as a *latching circuit*.

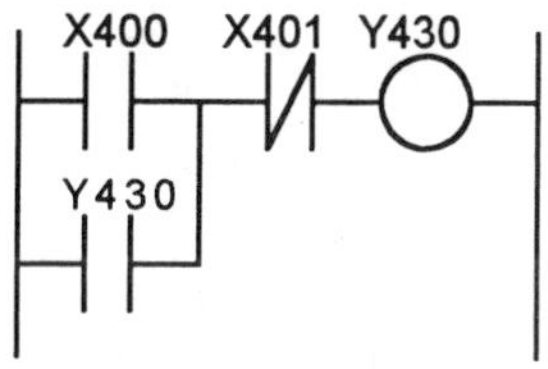

X400 X401 Y430 Y430

Figure 6.44 *Latching circuit*

In order to cope with events requiring time delays, so that events do not occur immediately a switch is closed but after some predetermined time, PLCs include *timers*. Thus if some output is to switch on after a delay we might have the ladder rung shown in Figure 6.45. The output is specified by T450, this being the type of notation used with Mitsubishi PLCs. The time delay is specified by the value given to K. This specified the number of units of the smallest time interval that is used by the timer. Thus if this is 0.1 s then with K = 100 we have a time delay of 10 s. Hence when switch X400 closes, there is a time delay of 10 s before the timer closes its normally open contacts. When these close there is then an output from Y430.

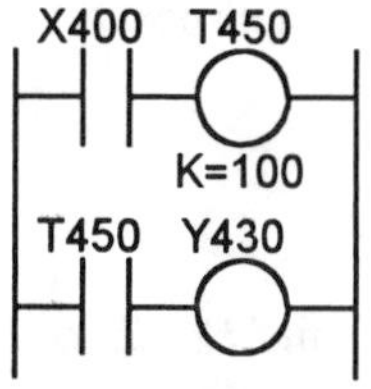

X400 T450 K=100 T450 Y430

Figure 6.45 *Use of a timer*

Each rung on the ladder represents a line in the program to be followed. To enter the program into the PLC a keyboard can be used with the keys having the graphic symbols for the ladder elements. Alternatively, each element on a ladder rung can be represented by letter code and the letters keyed in. This is referred to as a *statement or instruction list*. The statements/instructions used depend on the manufacturer of the PLC. Whatever form of input is used, the programming terminal translates the inputs into the binary code format which is necessary for the operation of the microcontroller.

As an example of a statement/instruction list, the following are some of the codes used with Mitsubishi PLCs.

LD	Start a rung of the ladder with an open contact.
LDI	Start a rung of the ladder with a closed contact.
...I	This is used in conjunction with other instructions, as illustrated above, to indicate the inverse.
AND	A series element and so an AND logic instruction.
OR	A parallel element and so an OR logic instruction.
K	Insert a constant
OUT	An output
END	End of the ladder and so end of the program.

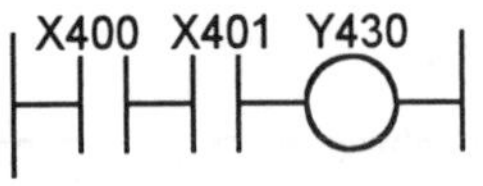

X400 X401 Y430

Figure 6.46 *An AND logic gate*

To illustrate their use, for the rung in Figure 6.46 we would use:

```
LD   X400
AND  X401
OUT  Y430
```

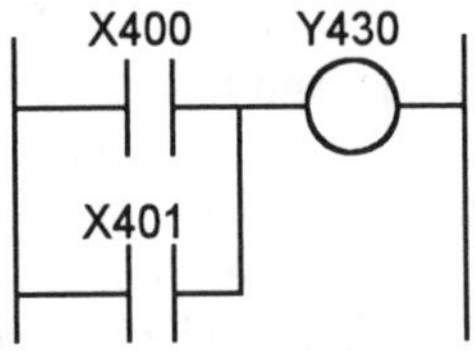

X400 Y430

X401

Figure 6.47 *An OR logic gate*

For the rung in Figure 6.47 we would use:

```
LD   X400
OR   X401
OUT  Y430
```

A program will generally consist of a number of rungs, each rung corresponding to an event in a controlled sequence. The above is just an indication of how programmable logic controllers can be used. In addition to the contacts, timers and outputs indicated above, they also other elements, such as counters and markers, which can be incorporated into programs.

Problems

Questions 1 to 20 have four answer options: A, B, C and D. Choose the correct answer from the answer options.

1 Decide whether each of these statements is True (T) or False (F).

An on-off temperature controller must have:
(i) An input of the error signal which switches the controller on or off.
(ii) An output signal which switches on or off the correction element.

A (i) T (ii) T
B (i) T (ii) F
C (i) F (ii) T
D (i) F (ii) F

2 Decide whether each of these statements is True (T) or False (F).

Oscillations of the variable being controlled occur with on–off temperature controller because:
(i) There is a time delay in switching off the correction element when the variable reaches the set value.
(ii) There is a time delay in switching on the correction element when the variable falls below the set value.

A (i) T (ii) T
B (i) T (ii) F
C (i) F (ii) T
D (i) F (ii) F

3 Decide whether each of these statements is True (T) or False (F).

With a proportional controller:
(i) The controller output is proportional to the error.
(ii) The controller gain is proportional to the error.

A (i) T (ii) T
B (i) T (ii) F
C (i) F (ii) T
D (i) F (ii) F

4 A steady state error will not occur when there is a change to the set value with a control system operating in the mode:

A Proportional
B Proportional plus derivative
C Derivative
D Proportional plus integral

Questions 5 to 8 concern the error input to a controller shown in Figure 6.48(a) and the possible controller outputs shown in Figure 6.48(b).

5 Which one of the outputs could be given by a proportional controller?
6 Which one of the outputs could be given by a derivative controller?
7 Which one of the outputs could be given by an integral controller?
8 Which one of the outputs could be given by a proportional plus integral controller?

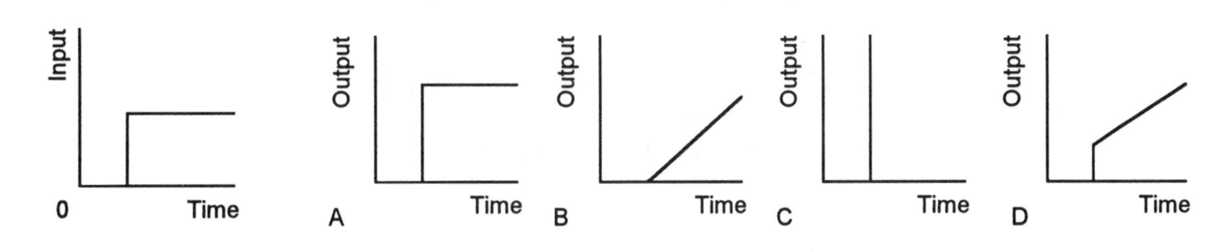

Figure 6.48 *(a) Error input to a controller, (b) possible controller outputs*

9 Decide whether each of these statements is True (T) or False (F).

With a controller operating with a derivative control mode:
(i) The controller output will be constant when the error is changing at a constant rate.
(ii) The controller output will be zero when there error is constant.

A (i) T (ii) T
B (i) T (ii) F
C (i) F (ii) T
D (i) F (ii) F

10 Decide whether each of these statements is True (T) or False (F).

With a controller operating in integral mode:
(i) The controller output will be zero when the error changes at a constant rate.
(ii) The controller output will increase at a constant rate when the error is constant.

A (i) T (ii) T
B (i) T (ii) F
C (i) F (ii) T
D (i) F (ii) F

11 With a ladder diagram, the symbol || when appearing on a rung represents:

A A capacitor.
B An output device.
C A pair of contacts normally open.
D A break in the program.

12 With a ladder diagram, the symbol O when appearing on a rung represents:

A An output device.
B A closed switch.
C An OR gate.
D An AND gate.

13 With a ladder diagram, the vertical lines between which the rungs are drawn represent:

A The PLC container.
B The edges of a microcontroller chip.
C The power rails.
D The input and output ports of the PLC.

Figure 6.49 *Ladder diagram*

14 For the rung of the lader diagram shown in Figure 6.49, the two sets of switch contacts are operating as:

A An OR logic system.
B An AND logic system.
C A NOR logic system.
D A NAND logic system.

Figure 6.50 *Ladder diagram*

15 For the rung of the ladder diagram shown in Figure 6.50, the switch contacts are operating as:

A An OR logic system.
B An AND logic system.
C A NOR logic system.
D A NAND logic system.

16 Decide whether each of these statements is True (T) or False (F).

With a PID process control system tested at startup using the ultimate cycle method with the derivative mode turned off and the integral mode set to its lowest settting, the period of oscillation was found to be 10 minutes with a proportional gain setting of 2. The optimum settings, using the criteria of Ziegler and Nicholls, will be:
(i) An integral constant of 0.2 min^{-1}.
(ii) A proportional gain setting of 1.2.

A (i) T (ii) T
B (i) T (ii) F
C (i) F (ii) T
D (i) F (ii) F

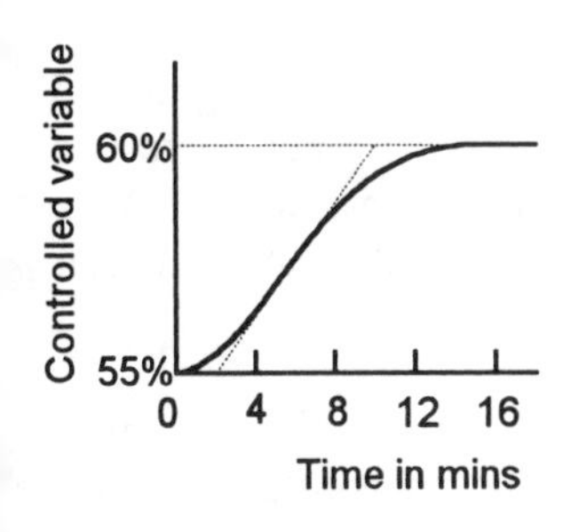

Figure 6.51 *Response graph*

17 Decide whether each of these statements is True (T) or False (F).

With a PI process constrol system tested by the process reaction method and the controller output changed by 10%, the response graph obtained was as shown in Figure 6.51. The optimum settings, using the criteria of Ziegler and Nicholls, will be:
(i) Proportional gain constant 7.2.
(ii) Integral constant 0.15 min^{-1}.

A (i) T (ii) T
B (i) T (ii) F
C (i) F (ii) T
D (i) F (ii) F

18 Decide whether each of these statements is True (T) or False (F).

With a PI process control system tested at startup using the ultimate cycle method with the derivative mode turned off and the integral mode set to its lowest settting, the period of oscillation was found to be 20 minutes with a proportional gain setting of 1.2. The optimum settings, using the criteria of Ziegler and Nicholls, will be:

(i) An integral constant of 0.06 min^{-1}.
(ii) A proportional gain setting of 1.2.

A (i) T (ii) T
B (i) T (ii) F
C (i) F (ii) T
D (i) F (ii) F

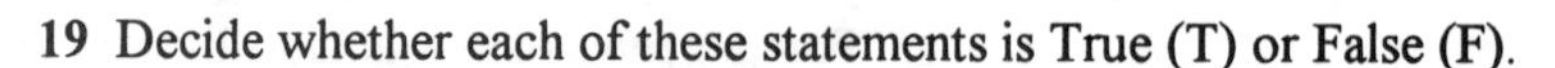

19 Decide whether each of these statements is True (T) or False (F).

For the operational amplifier circuit shown in Figure 6.52:
(i) The current through resistance 1 is virtually the same as that through resistance 2.
(ii) The operational amplifier is connected as a non-inverting amplifer.

A (i) T (ii) T
B (i) T (ii) F
C (i) F (ii) T
D (i) F (ii) F

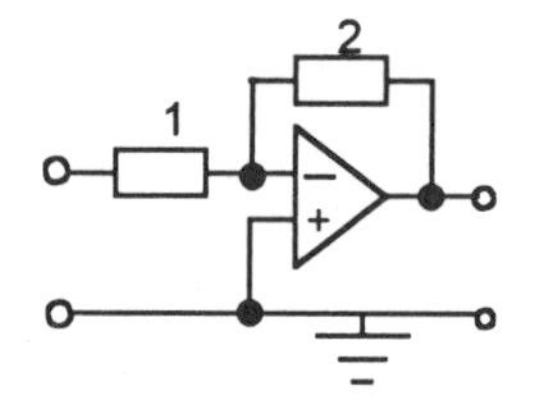

Figure 6.52 *Operational amplifier circuit*

20 Decide whether each of these statements is True (T) or False (F).

An element of a pnematic controller is the flapper–nozzle. For the flapper–nozzle shown in Figure 6.53, with a constant supply pressure:
(i) Increasing the separation of the flapper from the nozzle increases the output pressure.
(ii) The output presssure is a magnified version of the input pressure.

A (i) T (ii) T
B (i) T (ii) F
C (i) F (ii) T
D (i) F (ii) F

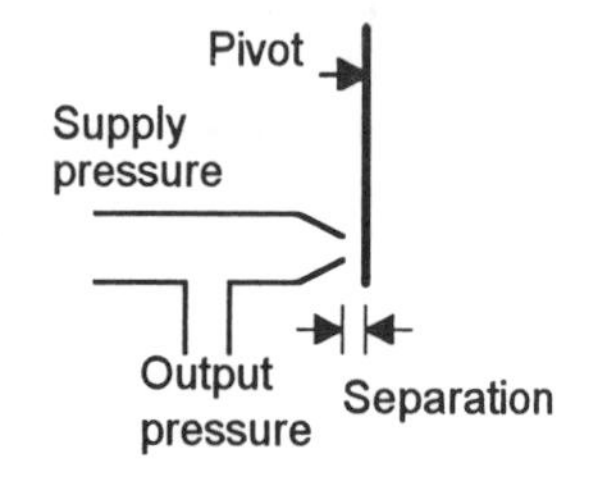

Figure 6.53 *Flapper–nozzle*

21 Describe the different control modes of proportional, integral and derivative and how the controller output is affected when there is a change in the error input signal and (a) the proportional mode, (b) the proportional plus integral, (c) the proportional plus integral plus derivative modes are used.

22 Expalin how a PID controller can be tuned to give the optiumum settings for the proportional constant, the integral constant and the derivative constant.

23 Describe examples of the following types of controllers, (a) lever, (b) flapper and nozzle, (c) operational amplifier.

7 Final control elements

7.1 Correction elements

The *correction element* or *final control element* is the element in a control system which is responsible for transforming the output of a controller into a change in the process which aims to correct the change in the controlled variable. Thus, for example, it might be a valve which is operated by the output from the controller and used to change the rate at which liquid passes along a pipe and so change the controlled level of the liquid in a cistern. It might be a motor which takes the electrical output from the controller and transforms it a rotatory motion in order to move a load and so control its position. It might be a switch which is operated by the controller and so used to switch on a heater to control temperature.

The term *actuator* is used for the part of a correction/final control element that provides the power, i.e. the bit which moves, grips or applies forces to an object, to carry out the control action. Thus a valve might have an input from the controller and be used to vary the flow of a fluid along a pipe and so make a piston move in a cylinder and result in linear motion. The piston/cylinder system is termed an actuator.

In this chapter pneumatic/hydraulic and electric correction/final control systems, along with actuators, are discussed.

7.2 Pneumatic and hydraulic systems

Process control systems frequently require control of the flow of a fluid. The valves used as the correction/final control elements in such situations are frequently pneumatically operated, even when the control system is otherwise electrical. This is because such pneumatic devices tend to be cheaper and more easily capable of controlling large rates of flow. The main drawback with pneumatic systems is, however, the compressibility of air. This makes it necessary to have a storage reservoir to avoid changes in pressure occurring as a result of loads being applied. Hydraulic signals do not have this problem and can be used for even higher power control devices. They are, however, expensive and there are hazards associated with oil leaks which do not occur with air leaks.

7.2.1 Current to pressure converter

Generally the signals required by a pneumatic final control element are in the region of 20 to 100 kPa gauge pressure, i.e. pressure above the atmospheric pressure. Figure 7.1 shows the principle of one form of a *current to pressure converter* (note that a pressure to current converter was discussed in section 2.4.8) that can be used to convert a current output from a controller, typically in the range 4 to 20 mA, to a pneumatic pressure signal of 20 to 100 kPa to operate a final control element. The current from the controller passes through coils mounted on a pivoted beam. As a consequence, the coils are then attracted towards a magnet, the

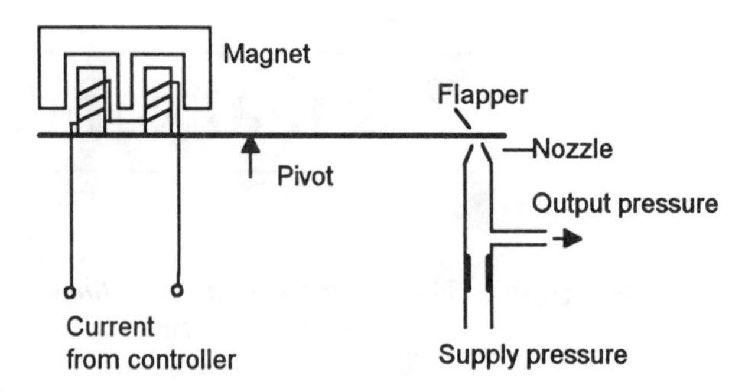

Figure 7.1 *Current to pressure converter*

extent of the attraction depending on the size of the current. The movement of the coils cause the lever to rotate about its pivot and so change the separation of a flapper from a nozzle. The position of the flapper in relation to the nozzle determines the size of the output pressure in the system (see section 6.8.1 for a discussion of the flapper–nozzle system).

7.2.2 Pressure sources

With a pneumatic system a source of pressurised air is required. This can be provided by an electric motor driving an air compressor (Figure 7.2). The air is drawn from the atmosphere via a filter. Since the air compressor increases the temperature of the air, a cooling system is likely to follow and, since air also contains a significant amount of moisture, a moisture separator to remove the moisture from the air. A storage reservoir is used to smooth out any pressure fluctuations due to the compressibility of air. A pressure relief valve provides protection against the pressure in the system rising above a safe level.

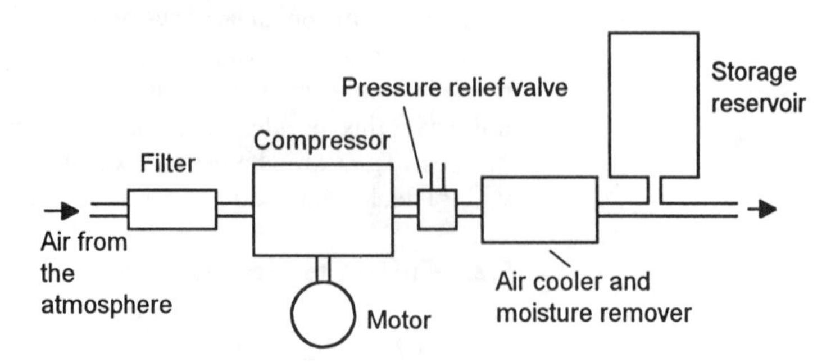

Figure 7.2 *A pressurised air source*

With a hydraulic system a source of pressurised oil is required. This can be provided by a pump driven by an electric motor. The pump pumps oil from a sump through a non-return valve and an accumulator and back to the sump (Figure 7.3). The non-return valve is to prevent the oil being back-driven to the pump. A pressure relief valve is included so that the pressure is released if it rises above a safe level. The accumulator is essentially just a container in which the oil is held under pressure against an external force and is there to smooth out any short-term fluctuations in the output oil pressure. If the oil pressure rises then the piston moves to increase the volume the oil can occupy and so reduces the pressure. If the oil pressure falls then the piston moves in to reduce the volume occupied by the oil and so increase its pressure.

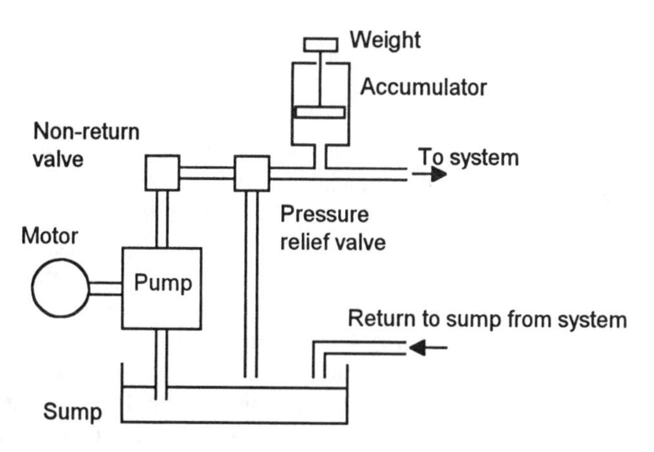

Figure 7.3 *A source of pressurised oil*

7.2.3 Control valves

Pneumatic and hydraulic systems use control valves to give direction to the flow of fluid through a system, control its pressure and control the rate of flow. These types of valve can be termed *directional control valves*, *pressure control valves* and *flow control valves*. Directional control valves, sometimes termed *finite position valves* because they are either completely open or completely closed, i.e. they are on/off devices, are used to direct fluid along one path or another. They are equivalent to electric switches which are either on or off. Pressure control valves, often termed pressure regulator valves, react to changes in pressure in switching a flow on or off, or varying it. Flow control valves, sometimes termed *infinite position valves*, continuously vary the rate at which a fluid passes through a pipe. These three types of valve are discussed in more detail in sections 7.3, 7.4 and 7.5.

7.2.4 Actuators

As an illustration of the type of role played by actuators, consider the example illustrated in Figure 7.4. An electronic controller is used to provide a current which is used to control a valve and hence the rate at which hydraulic fluid flows through a pipe. This is fed to an actuator which converts the rate of flow of hydraulic fluid into a linear motion.

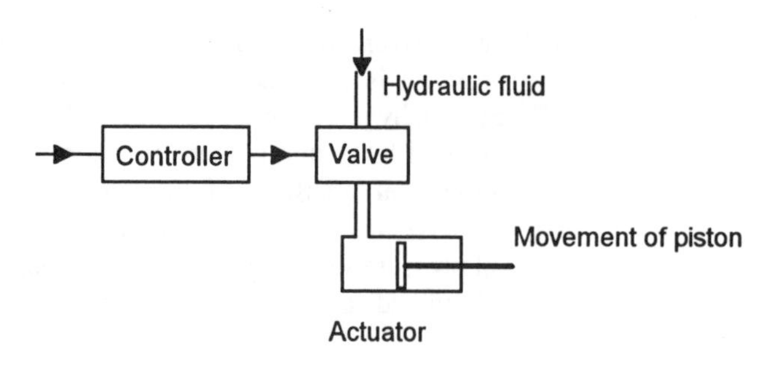

Figure 7.4 *Illustration of role of an actuator*

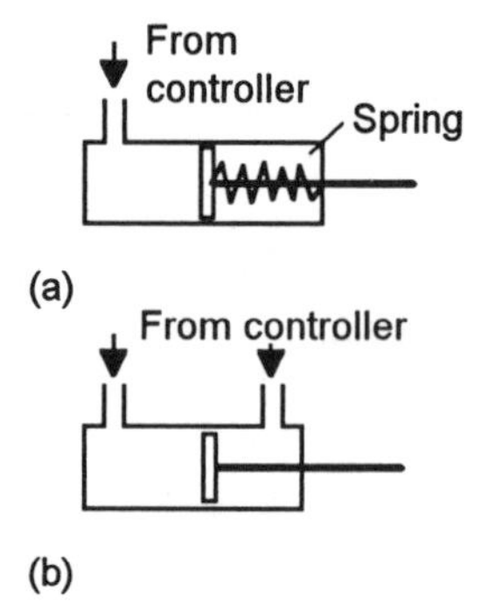

Figure 7.5 *(a) Single acting, (b) double acting cylinder*

The *hydraulic* or *pneumatic cylinder* is an example of a linear actuator, the principles and form being the same for both versions with the differences being purely a matter of size as a consequence of the higher pressures used with hydraulics. The hydraulic/pneumatic cylinder consists of a hollow cylindrical tube along which a piston can slide. Figure 7.5(a) shows the single acting form and Figure 7.5(b) the double acting form. The *single acting* form has the control pressure applied to just one side of the piston, a spring often being used to provide the opposition to the movement of the piston. The piston can only be moved in one direction along the cylinder by the signal from the controller. The *double acting* form has control pressures that can be applied to each side of the piston. When there is a difference in pressure between the two sides the piston moves, the piston being able to move in either direction along the cylinder.

The choice of cylinder is determined by the force required to move the load and the speed required. Hydraulic cylinders are capable of much larger forces than pneumatic cylinders. However, pneumatic cylinders are capable of greater speeds.

Since pressure is force per unit area, the force produced by a piston in a cylinder is equal to the cross-sectional area of the piston, this being effectively the same as the internal cross-sectional area of the cylinder, multiplied by the difference in pressure between the two sides of the piston. Thus for a pneumatic cylinder with a pressure difference of 500 kPa and having an internal diameter of 50 mm,

$$\text{force} = \text{pressure} \times \text{area} = 500 \times 10^3 \times \tfrac{1}{4}\pi \times 0.050^2 = 982 \text{ N}$$

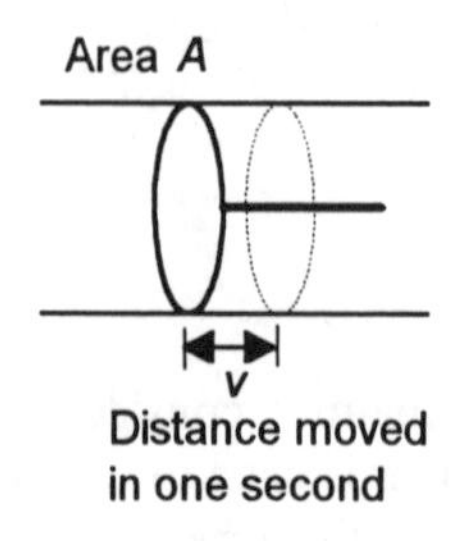

Figure 7.6 *Movement of piston*

A hydraulic cylinder with the same diameter and a pressure difference of 15 000 kPa, hydraulic cylinders being able to operate with higher pressures than pneumatic cylinders, will give a force of 29.5 kN. Note that the maximum force available is not related to the flow rate of hydraulic fluid or air into a cylinder but is determined solely by the pressure and piston area.

The speed with which the piston moves in a cylinder is determined by the rate at which fluid enters the cylinder. If the flow rate of hydraulic liquid into a cylinder is a volume of Q per second, then the piston must sweep out a volume of Q. If a piston moves with a velocity v then, in one second, it moves a distance of v (Figure 7.6). But for a piston of cross-sectional area

A this must mean that the volume swept out by the piston in 1 s is Av. Thus we must have:

$$Q = Av$$

Thus the speed v of a hydraulic cylinder is equal to the flow rate of liquid Q through the cylinder divided by the cross-sectional area A of the cylinder. The speed is determined by just the piston area and the flow rate. For example, for a hydraulic cylinder of diameter 50 mm and a hydraulic fluid flow of 7.5×10^{-3} m^3/s:

$$\text{speed } v = \frac{Q}{A} = \frac{7.5 \times 10^{-3}}{\frac{1}{4}\pi \times 0.050^2} = 3.8 \text{ m/s}$$

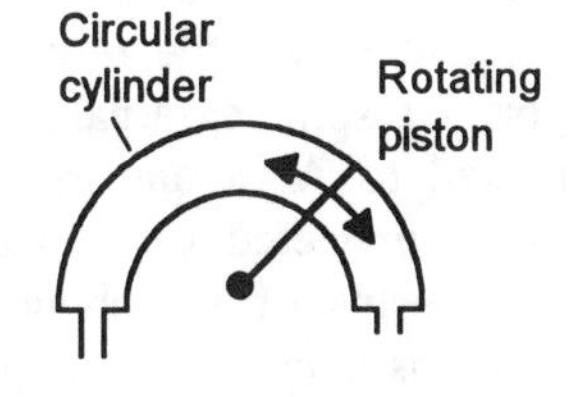

Figure 7.7 *Rotary actuator*

The hydraulic/pneumatic cylinder is termed a linear actuator since the result of the application of pressure is a linear motion. Rotary actuators give rotary motion as a result of the applied fluid pressure. Figure 7.7 shows a rotary actuator which gives partial rotary movement. Continuous rotation is possible with some forms and then they are the equivalent of electric motors. Figure 7.8 shows one form, known as a *vane motor*. The vanes are held out against the walls of the motor by springs or hydraulic pressure. Thus, when there is a pressure difference between the inlet and outlets of the motor, rotation occurs.

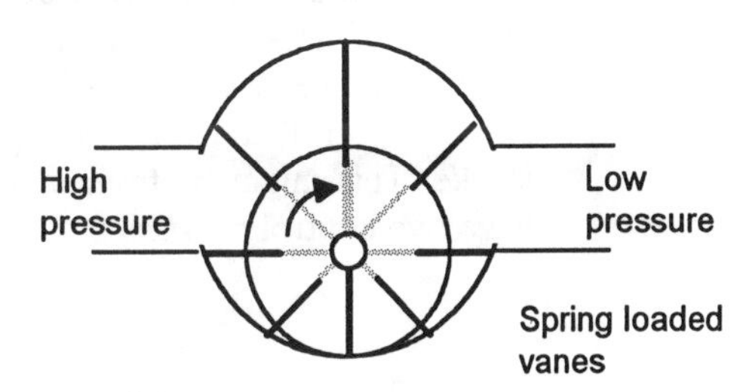

Figure 7.8 *Vane motor*

Example

A hydraulic cylinder is to be used in a manufacturing operation to move a workpiece through a distance of 250 mm in 20 s. If a force of 50 kN is required to move the workpiece, what is the required pressure difference and hydraulic liquid flow rate if a cylinder with a piston diameter of 150 mm is to be used?

As derived above, the force produced by the cylinder is equal to the product of the cross-sectional area of the cylinder and the working pressure. Thus the required pressure is:

$$\text{pressure} = \frac{F}{A} = \frac{50 \times 10^3}{\frac{1}{4}\pi \times 0.150^2} = 2.8 \times 10^6 \text{ Pa} = 2.8 \text{ MPa}$$

The average speed required is 250/20 = 12.5 mm/s. As derived above, the speed of a hydraulic cylinder is equal to the flow rate of liquid through the cylinder divided by the cross-sectional area of the cylinder. Thus the required flow rate is:

$$\text{flow rate} = \text{speed} \times \text{area} = 0.0125 \times \tfrac{1}{4}\pi \times 0.150^2 = 2.2 \times 10^{-4}\ \text{m}^3/\text{s}$$

7.3 Directional control valves

Directional control valves are widely used in control systems as elements for switching on or off hydraulic or pneumatic pressures which can then, via some actuator, control the movement of some item. Figure 7.9 shows a basic, finite position, control valve system with four ports. The input and output connections to valves are through, what are termed, ***ports***. The actuator, in this example a double acting cylinder, is connected to the ports A and B. The pressure supply from the pump or compressor is connected to port P and, in the case of a hydraulic valve, the fluid is returned to the tank through port T. In the case of a pneumatic system valve, the return air would be vented to the atmosphere from this port. By some external means we can have P connected to B and T connected to A in one instance (Figure 7.9(a)) and then have it switched to P connected to A and T connected to B (Figure 7.9(b)). This external means of switching the connections might be a mechanical lever or perhaps a current through a solenoid. In Figure 7.9(a), the pressure supply through port P is connected to port B and A to the vent port T and drives the piston in the cylinder to one end of its stroke. This might be termed the extend stroke. In Figure 7.9(b), the pressure supply through port P is connected to port A and B to the vent port T and drives the piston in the cylinder to the other end of its stroke. This might be termed the retract stroke. The valve is then said to have two control *positions*.

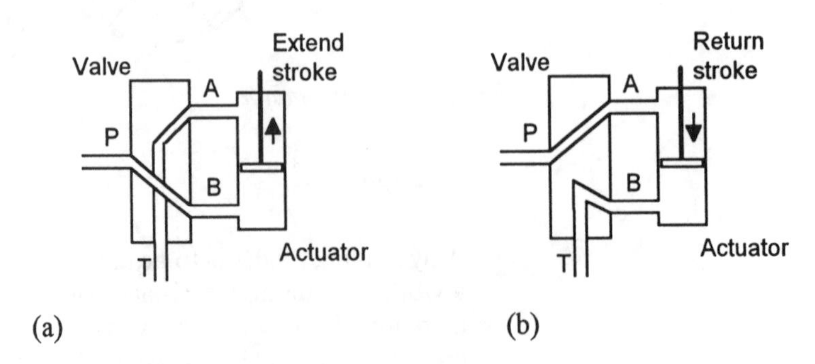

Figure 7.9 *A 4 port directional control valve*

Directional valves are specified in terms of the number of ports and number of control positions they have. Thus the valve shown in Figure 7.9 is one with four ports and two positions, this being termed a 4/2 valve. The two positions are (a) and (b) in Figure 7.9. The symbol used for control valves consists of a square for each of its switching positions. Thus the 4/2

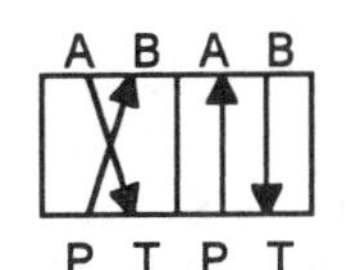

Figure 7.10 *A 4/2 valve*

valve will have two squares. Arrow-headed lines are used to indicate the directions of flow in each of the positions. The ports are labelled P for pressure supply, R, S and T for the exhaust/return ports with T normally used for the hydraulic return port and R for a pneumatic exhaust port, and A, B, etc. for the working pressure lines. Figure 7.10 shows the 4/2 valve described in Figure 7.9. The first square indicates one position for the valve, the piston extended state (Figure 7.9(a)). With this position, the pressure supply is connected via P to port B and port A is returning hydraulic fluid through port T. The second square indicates the other position for the valve, the retracted state (Figure 7.9(b)). Port P is now supplying fluid to port A and port B is now returning fluid through port T.

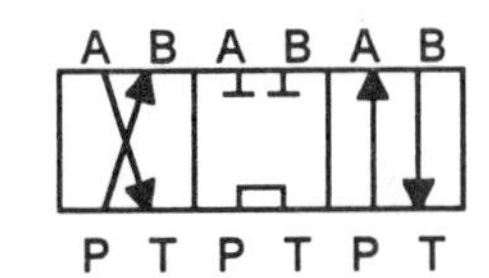

Figure 7.11 *A 4/3 valve*

A 4/3 valve is one with four ports and three positions. These three positions can be extend, off and retract. The off position is when the piston in the actuator is held in the centre of the cylinder. Figure 7.11 indicates the possible valve actions for these positions. The first position shown is the extend position, as in the 4/2 valve above. The second position is the off position. With the 4/3 valve in the off position, ports P and T are connected and ports A and B are blocked. The third position is the retract position, as in the 4/2 valve above.

The valve symbols shown in Figures 7.10 and 7.11 do not show how the valve is made to switch between its various positions. Figure 7.12 shows some of the symbols which are used to indicate the various ways the valves can be operated. More than one of these symbols might be used with a particular valve symbol.

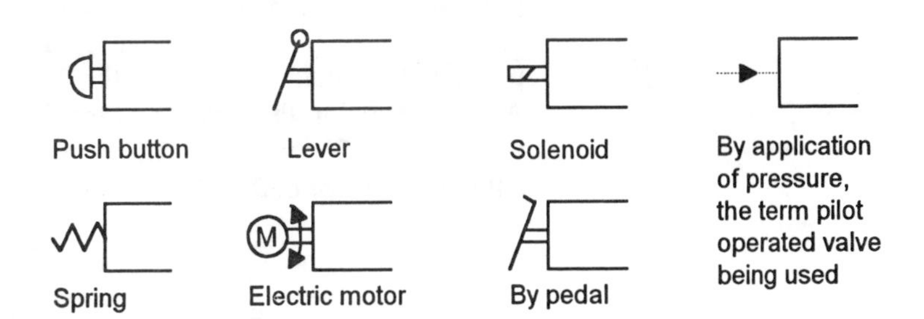

Figure 7.12 *Valve operation symbols*

To illustrate the use of such symbols, Figure 7.13 shows a 4/2 valve. When the push-button is depressed the piston extends. The push-button movement gives the state indicated by the symbols used in the square to which it is attached, in this case the first square. When the push-button is released, the spring pushes the valve back to its initial position and the piston retracts. The spring movement gives the state indicated by the symbols used in the square to which it is attached, i.e. the second square. This second square then gives the normal state of the valve.

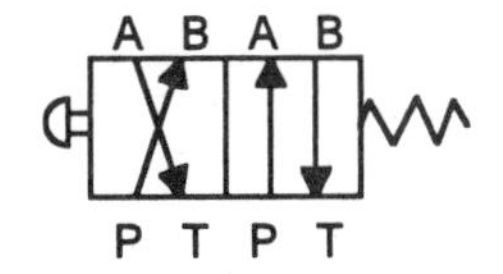

Figure 7.13 *4/2 valve*

To illustrate the use of such symbols and indicate a simple pneumatic control system, Figure 7.14 shows a lift. Two push-button 2/2 valves are used. When the button on the up valve is pressed, the air pressure supply is connected via ports P and A to the cylinder and the load is lifted. When the up button is released, the spring causes ports P and A to become closed and so the load remains in its lifted position. When the button on the down valve is pressed, the air pressure is vented to the atmosphere since port A is now connected to port R which is a vent (note the symbol, an open arrow head, for a vent to the atmosphere) and the load is lowered. When the down button is released, the spring causes ports R and A to become closed and so the load remains in its lowered position.

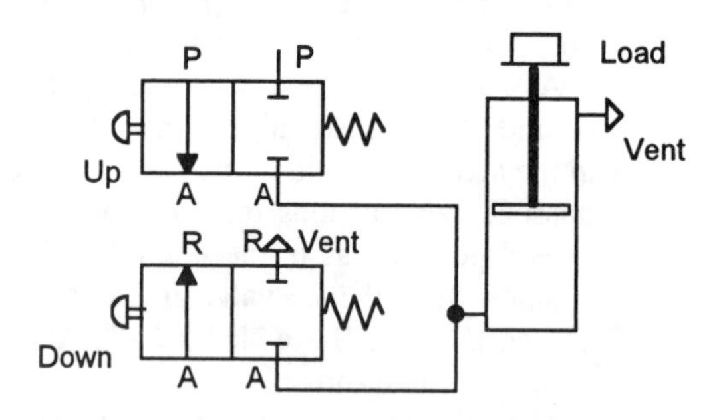

Figure 7.14 *Pneumatic lift system*

Another simple example is for a door to open if either push-button A on one side of the door is pressed or push-button B on the other side is pressed. Figure 7.15 shows a possible pneumatic circuit. When button 1 is pressed the air pressure is connected via port P to A and so moves the piston in the cylinder and opens the door. When the button is released, port A is connected to the vent R and the door closes. A similar situation occurs when button 2 is pressed. The arrangement is an OR logic gate in that either button 1 or button 2 can open the door.

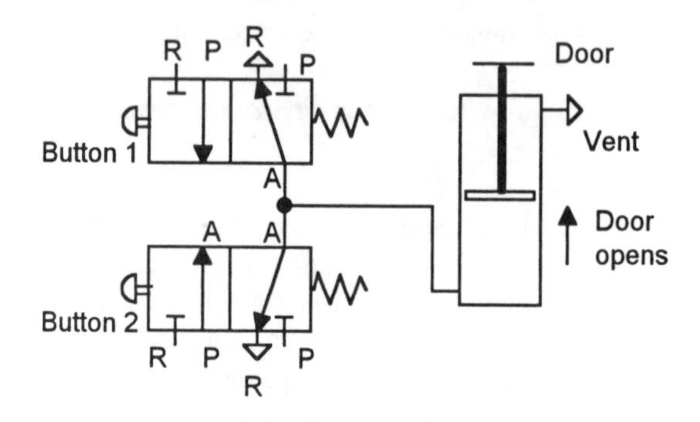

Figure 7.15 *The door-opening circuit*

7.3.1 Forms of directional control valves

Two basic forms of directional control valves are poppet valves and spool valves. The following indicates the basic structural details.

Figure 7.16 shows the basic form of a simple *2/2 poppet valve*. In poppet valves, balls, discs or cones are used in conjunction with valve seats to control the flow. In the figure a ball is shown. The valve is normally in the closed condition. When the push button is depressed, the ball is pushed out of its seat and flow can occur in the gap between the ball and the walls of the valve. When the button is released, the spring forces the ball back up against its seat and so closes off the flow.

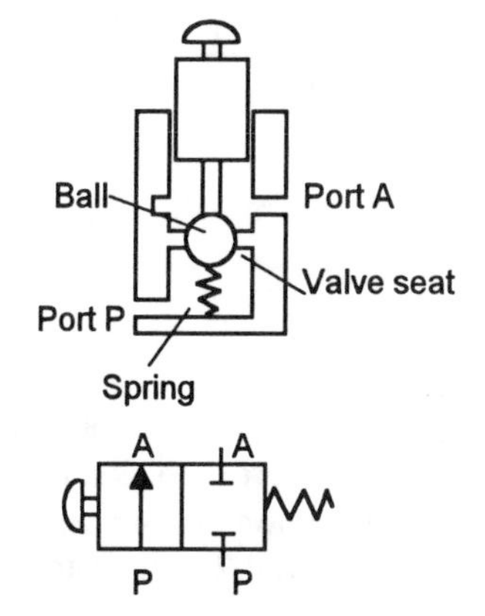

Figure 7.16 *A 2/2 poppet valve and its symbol*

Figure 7.17 shows the basic form of a *3/2 poppet valve*. With the push-button released, ports A and R are connected and port P is closed. When the push-button is pressed, the valve stem pushes down to close port R and push the valve disc down, so opening port P. Thus port R is closed and ports A and P connected. When the push-button is released, the spring returns the valve to its initial state.

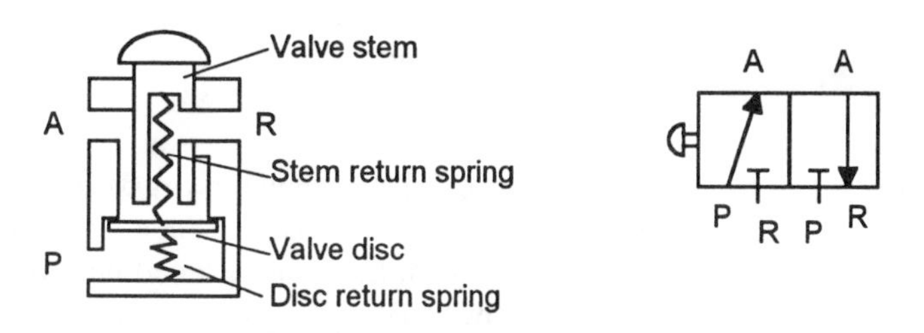

Figure 7.17 *A 3/2 poppet valve and its symbol*

Another type of finite position control valve is the *spool valve*. Such valves have spools, or sliders, moving within the valve body. Figure 7.18 shows the form for a 4/2 valve. As shown, port A is connected to P with B and T both closed. When the spool is moved to the left, by perhaps a push button or some other mechanism, then B is connected to P with A and T now closed.

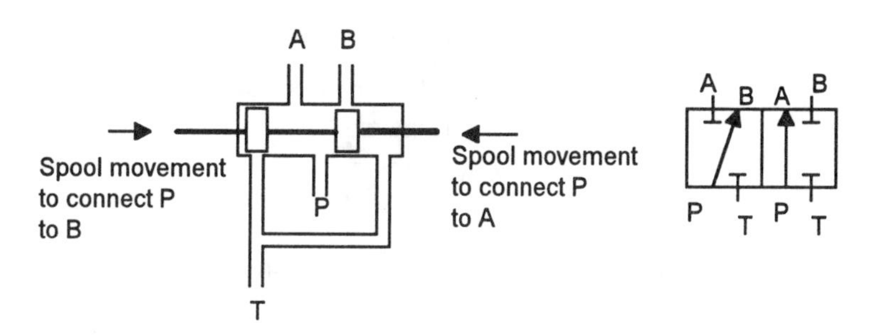

Figure 7.18 *A two-way spool valve and its symbol*

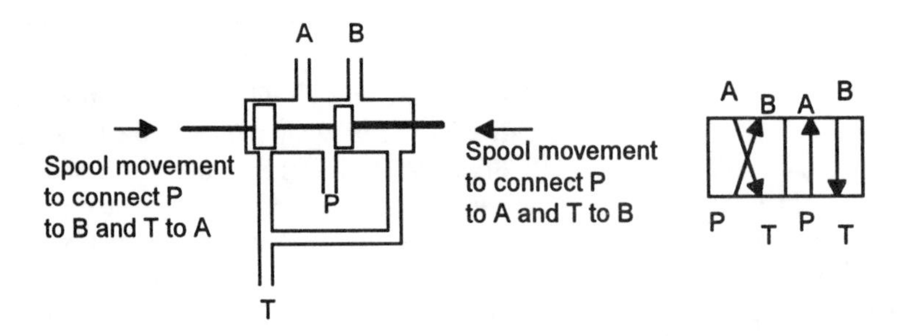

Figure 7.19 *A four-way spool valve and its symbol*

The valve shown in Figure 7.18 has just two possible paths for air or fluid to flow through the valve, P to A or P to B. It is thus referred to as a two-way shuttle valve. Figure 7.19 shows how, by a modification of the position of the spools in the spool valve shown in Figure 7.18, a 4/2 valve can be produced for which there are four possible paths through it.

In a similar way to the shuttle valves described above, *rotary spool valves* have a rotating spool which, when it rotates, opens and closes ports in a similar way.

7.4 Pressure control valves

Figure 7.20(a) shows a *pressure relief valve* which is normally closed so that air cannot escape from a pneumatic system. However, if the pressure in the system rises above a certain level, the force resulting from the pressure on the ball can overcome the force exerted by the spring with the result that the valve opens and permits air to escape and so reduce the pressure in the system. The pressure at which this occurs can be adjusted by screwing up or down the cap on the valve. This valve can thus be used as a pressure relief valve to safeguard a system against excessive pressures. Similar forms of valve can be used with hydraulic systems, the only real difference being that the escaping fluid is returned to the sump rather than being vented to the atmosphere. The symbol for a pressure relief valve is shown in Figure 7.20(b), this indicating that the opposing force is by a spring and that the exhaust port is to a reservoir or to the atmosphere.

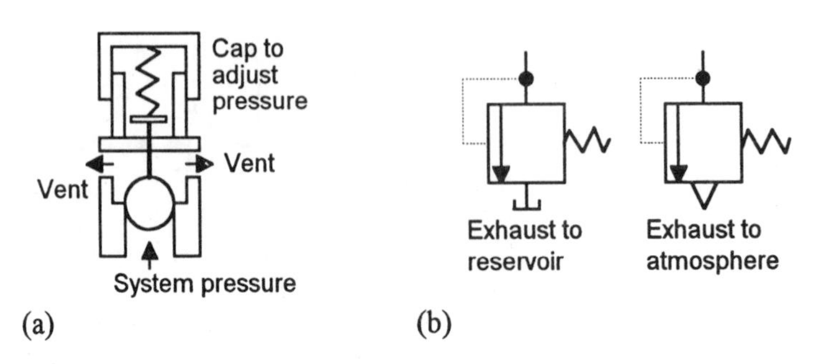

Figure 7.20 *Pressure relief valve and its symbols*

Another form of pressure control valve is one which is used to regulate the pressure and maintain it at a constant value. Figure 7.21 shows such a *pressure regulator* and the symbol used in circuit diagrams. The outlet pressure, i.e. the pressure in the system, is sensed by the diaphragm. If the pressure is too low, the spring forces the diaphragm downwards and moves the poppet down. This admits more air into the system and so raises the pressure. If the pressure is too high, the diaphragm is forced upwards and moves the poppet upwards. This cuts off the air into the system. When the pressure is just right, the force on the diaphragm due to the pressure just balances the force applied by the spring. The air entering the system is then just right to maintain the required pressure.

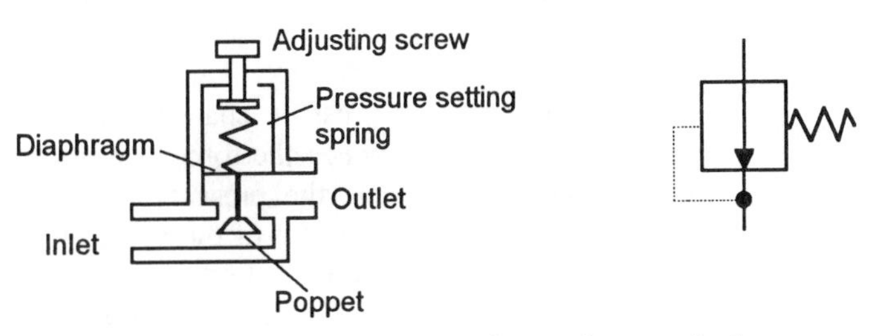

Figure 7.21 *Pressure regulator valve and its symbol*

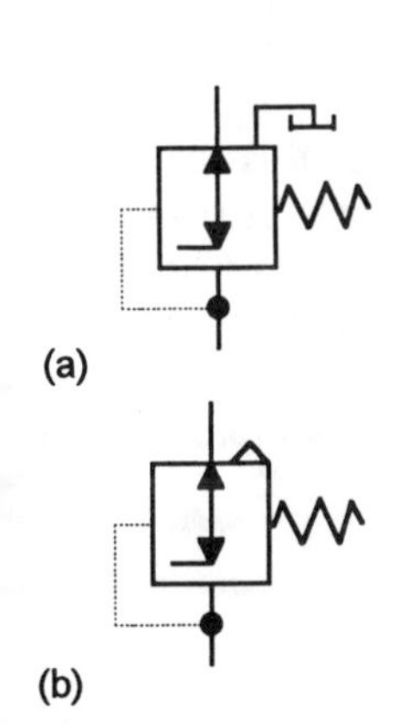

Figure 7.22 *Relieving pressure regulator, relief to (a) reservoir, (b) atmosphere*

The above regulator valve is termed a *non-relieving regulator* since when the pressure is too high there is no relief of the pressure by venting to the atmosphere but just a reliance on natural leakage in the system bringing the pressure back down to the required level. A valve where there is relief when the pressure becomes too high is termed a *relieving pressure regulator*, Figure 7.22 showing the symbols, (a) being for hydraulic pressure systems when the hydraulic fluid is returned to a reservoir and (b) for a pneumatic system where the air is vented to the atmosphere.

7.5 Flow control valves

In many control systems the rate of flow of a fluid along a pipe is controlled and the final control element used is the *pneumatic control valve*. These valves involve pneumatic action to move a variable restriction in the flow path. The plug valve moves a plug into the flow path, Figure 7.23 showing the basic form of such a valve. Movement of the valve stem causes the plug to move and so alter the size of the gap through which the fluid can flow. With the pneumatic control valve, this movement of the stem results from the use of a diaphragm moving against a spring and controlled by air pressure. Figure 7.24 shows the forms known as direct and reverse action valve actuators. The air pressure from the controller exerts a force on one side of the diaphragm, the other side of the diaphragm being at atmospheric pressure, which is opposed by the force due to the spring on the other side.

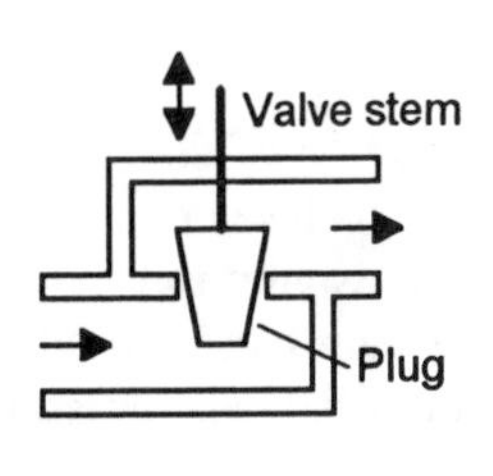

Figure 7.23 *The plug valve*

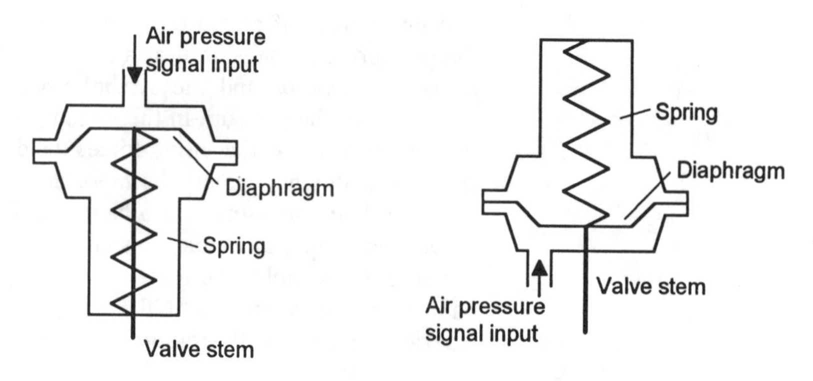

Figure 7.24 *(a) Direct, (b) reverse action valve actuators*

When the air pressure changes then the diaphragm moves until there is equilibrium between the forces resulting from the pressure and those from the spring. Thus the pressure signals from the controller result in the movement of the stem of the valve. The difference between the direct and reverse forms is the position of the spring.

The force F acting on the diaphragm due to the pressure is the gauge pressure p, i.e. the difference between the control pressure and the atmospheric pressure, multiplied by the diaphragm area A. The force provided by the spring, when the stem of the valve moves through a distance x, is given by $F = kx$ with k being a constant, assuming that the spring obeys Hooke's law. The diaphragm moves until the two forces are in equilibrium. Thus we have:

$$kx = pA$$

and the displacement x of the valve stem is proportional to the gauge pressure p.

The air pressure signals from the controller to give 0 to 100% correction generally vary from 0.2 to 1.0 bar (1 bar is 0.1 MPa) above the atmospheric pressure. Consider what diaphragm area is needed to 100% open a control valve if a force of 500 N must be applied to the valve stem when the input gauge pressure from the controller is 0.1 MPa. The force F applied to the diaphragm of area A by a pressure p is given by $p = F/A$. Hence the area $A = 500/(0.1 \times 10^6) = 0.005\ \text{m}^2$.

7.5.1 Valve bodies and plugs

There are many forms of valve body and plug. The selection of the form of body and plug determine the characteristic of the control valve, i.e. the relationship between the valve stem position and the flow rate through it. For example, Figure 7.25 shows how the selection of plug can be used to determine whether the valve closes when the controller air pressure increases or opens when it increases.

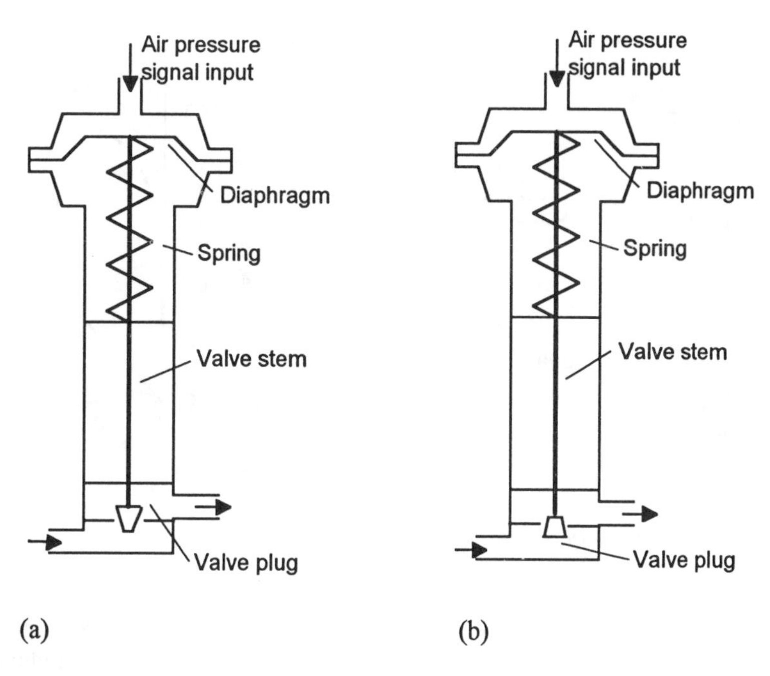

Figure 7.25 *Direct action: (a) air pressure increase to close, (b) air pressure increase to open*

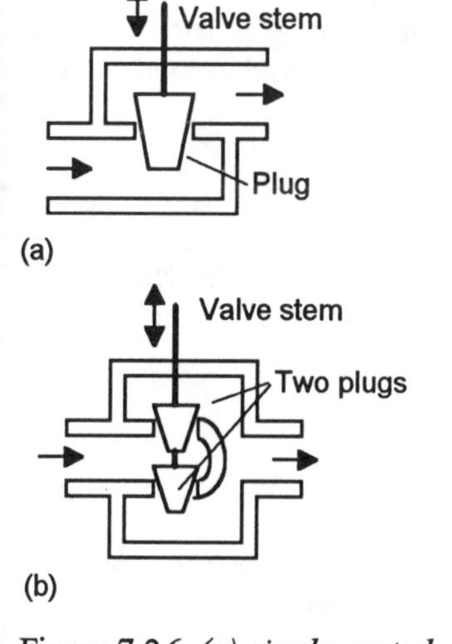

Figure 7.26 *(a) single seated, (b) double seated*

The term *single seated* is used for a valve where there is just one path for the fluid through the valve and so just one plug is needed to control the flow (Figure 7.26(a)). The term *double seated* is used for a valve where the fluid on entering the valve splits into two streams, each stream passing through an orifice controlled by a plug and thus two plugs with such a valve (Figure 7.26(b)). A single-seated valve has the advantage that is can be closed more tightly than a double seated one but the disadvantage that the force on the plug due to the flow is much higher and so the diaphragm has to exert considerably higher forces on the stem. This can result in problems in accurately positioning the plug and so double seated valves have an advantage.

The shape of the plug determines the relationship between the stem movement and the effect on the flow rate. Figure 7.27 shows three commonly used types:

1 Quick-opening
2 Linear
3 Equal percentage

and how the percentage volumetric rate of flow is related to the percentage displacement of the valve stem.

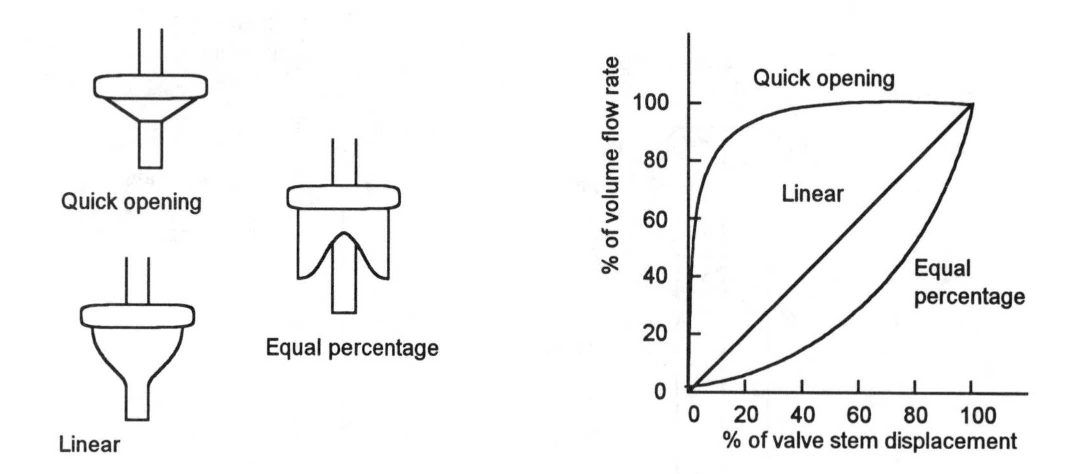

Figure 7.27 *Examples of plugs and their characteristics*

With the *quick-opening* type, the shape of the plug is such that a large change in flow rate occurs for a small movement of the valve stem. Such a plug is used where on/off control of flow rate is required.

With the *linear-contoured* type, the shape of the plug is such that the change in flow rate is proportional to the change in displacement of the valve stem, i.e.:

$$\text{change in flow rate} = k\,(\text{change in stem displacement})$$

where k is a constant. If Q is the flow rate at a valve stem displacement S and Q_{max} is the maximum flow rate at the maximum stem displacement S_{max}, then we have:

$$\frac{Q}{Q_{max}} = \frac{S}{S_{max}}$$

or the percentage change in the flow rate equals the percentage change in the stem displacement. Such valves are widely used for the control of liquids entering cisterns when the liquid level is being controlled.

With the *equal percentage* type of plug, the amount by which the flow rate changes ΔQ for a change in valve stem position ΔS is proportional to the value of the flow Q when the change occurs, i.e.:

$$\frac{\Delta Q}{\Delta S} \propto Q$$

Hence we can write

$$\frac{\Delta Q}{\Delta S} = kQ$$

where k is a constant. Generally this type of valve does not cut off completely when at the limit of its stem travel, thus when $S = 0$ we have $Q = Q_{min}$. If we write this expression for small changes and then integrate it we obtain:

$$\int_{Q_{min}}^{Q} \frac{1}{Q} dQ = k \int_{0}^{S} dS$$

Hence:

$$\ln Q - \ln Q_{min} = kS$$

If we consider the flow rate Q_{max} is given by S_{max} then:

$$\ln Q_{max} - \ln Q_{min} = kS_{max}$$

Eliminating k from these two equations gives:

$$\frac{\ln Q - \ln Q_{min}}{\ln Q_{max} - \ln Q_{min}} = \frac{S}{S_{max}}$$

$$\ln \frac{Q}{Q_{min}} = \frac{S}{S_{max}} \ln \frac{Q_{max}}{Q_{min}}$$

and so:

$$\frac{Q}{Q_{min}} = \left(\frac{Q_{max}}{Q_{min}}\right)^{S/S_{max}}$$

The term *rangeability* R is used for the ratio Q_{max}/Q_{min}.

Example

A valve has a stem movement at full travel of 30 mm and has a linear plug which has a minimum flow rate of 0 and a maximum flow rate of 20 m^3/s. What will be the flow rate when the stem movement is 15 mm?

The percentage change in the stem position from the zero setting is $(15/30) \times 100 = 50\%$. Since the percentage flow rate is the same as the percentage stem displacement, then a percentage stem displacement of 50% gives a percentage flow rate of 50%, i.e. 10 m^3/s.

Example

A valve has a stem movement at full travel of 30 mm and an equal percentage plug. This gives a flow rate of 2 m^3/s when the stem position is 0. When the stem is at full travel there is a maximum flow rate of 20 m^3/s. What will be the flow rate when the stem movement is 15 mm?

Using the equation:

$$\frac{Q}{Q_{\text{min}}} = \left(\frac{Q_{\text{max}}}{Q_{\text{min}}}\right)^{S/S_{\text{max}}}$$

$$\frac{Q}{2} = \left(\frac{20}{2}\right)^{15/30}$$

gives $Q = 6.3\ \text{m}^3/\text{s}$.

7.5.2 Rangeability, turndown and valve sizing

Rangeability is defined as the ratio of the maximum to minimum rates of controlled flow. Thus if the minimum controllable flow is 2.0% of the maximum controllable flow, then the rangeability is 100/2.0 = 50.

Valves are often not required to handle the maximum possible flow and the term *turndown* is used for the ratio:

$$\text{turndown} = \frac{\text{normal maximum flow}}{\text{minimum controllable flow}}$$

For example, a valve might be required to handle a maximum flow which is 70% of that possible. With a minimum flow rate of 2.0% of the maximum flow possible, then the turndown is 70/2.0 = 35.

The term *control valve sizing* is used for the procedure of determining the correct size, i.e. diameter, of the valve body. A control valve changes the flow rate by introducing a constriction in the flow path (Figure 7.28). But introducing such a constriction introduces a pressure difference between the two sides of the constriction. Application of Bernoulli's equation to a constriction gives:

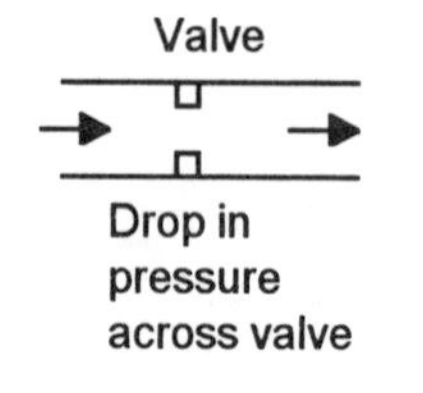

Figure 7.28 *Valve as a constriction*

$$\text{rate of flow } Q = K\sqrt{\text{pressure drop}}$$

where K is a constant which depends on the size of the constriction. The equations used for determining valve sizes are based on this equation. For a liquid, this equation is written as:

$$Q = A_V\sqrt{\frac{\Delta p}{\rho}}\ \text{m}^3/\text{s}$$

where A_V is the *valve flow coefficient*, Δp the pressure drop in Pa across the valve and ρ the density in kg/m^3 of the fluid. Because the equation was originally specified with pressure in pounds per square inch and flow rate in American gallons per minute, another coefficient C_V based on these units is widely quoted. With such a coefficient and the quantities in SI units, we have:

$$Q = 2.37 \times 10^{-5} C_V \sqrt{\frac{\Delta p}{\rho}} \text{ m}^3\text{/s}$$

or

$$Q = 0.75 \times 10^{-6} C_V \sqrt{\frac{\Delta p}{G}} \text{ m}^3\text{/s}$$

G is the specific gravity (relative density) and Δp is the pressure difference. Other similar equations are available for gases and steam. For gases:

$$Q = 6.15 \times 10^{-4} C_V \sqrt{\frac{\Delta p \times p}{TG}} \text{ mm}^3\text{/s}$$

where T is the temperature on the Kelvin scale and p the inlet pressure. For steam:

$$Q = 27.5 \times 10^{-6} C_V \sqrt{\frac{\Delta p}{V}} \text{ kg/s}$$

where V is the specific volume of the steam in m^3/kg, the specific volume being the volume occupied by 1 kg. Table 7.1 shows some typical values of A_V, C_V and the related valve sizes.

Table 7.1 Flow coefficients and valve size

Flow coefficients	Valve size in mm							
	480	640	800	960	1260	1600	1920	2560
C_V	8	14	22	30	50	75	110	200
$A_V \times 10^{-5}$	19	33	52	71	119	178	261	474

Example

Determine the valve size for a valve that is required to control the flow of water when the maximum flow rate required is 0.012 m^3/s and the permissible pressure drop across the valve at this flow rate is 300 kPa.

Rearranging the above equation and taking the density of water as 1000 kg/m^3,

$$A_V = Q\sqrt{\frac{\rho}{\Delta p}} = 0.012\sqrt{\frac{1000}{300 \times 10^3}} = 69.3 \times 10^{-5}$$

Thus, using Table 7.1, this value of coefficient indicates that the required valve size is 960 mm.

Figure 7.29 *Throttle valve symbol*

Figure 7.30 *Butterfly valve*

7.5.3 Throttle valves

The term *throttle valve* or *flow regulator* is used for a valve that restricts the flow of a fluid. Figure 7.29 shows the circuit symbol. Such a valve is used with car engines in a fuel injection system to control the rate at which air flows through to mix with the fuel. There it takes the form of a *butterfly valve*. Such a valve just consists of a large disc which is rotated inside the pipe (Figure 7.30) and is widely used as a throttling valve for gases.

Another form of throttling valve takes the form of a screw which is screwed in to an opening and moves a plug into a gap. Such a valve might be used in a pneumatic or hydraulic circuit to give speed control of a cylinder. Earlier in this chapter a circuit (Figure 7.15) was given for the opening of a door as a result of a push button being pressed on either side. While this circuit would work, there would be no control over the speed at which the door is opened or closed and, unless they are very quick, this could catch a person walking through the door. Figure 7.31 shows how the circuit could be modified by the introduction of a throttle valve to slow down the speed at which the pressure in the cylinder changes.

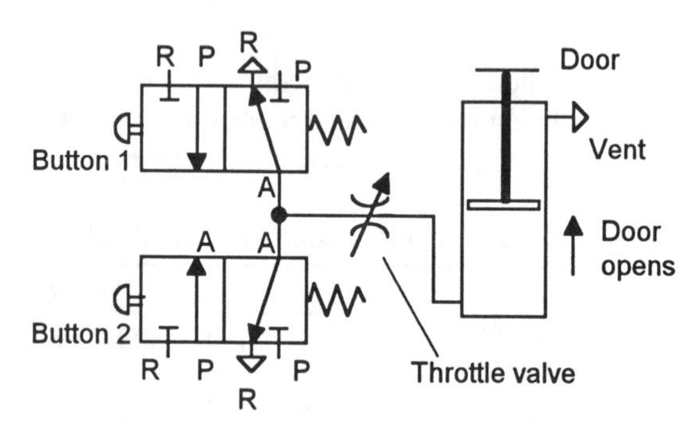

Figure 7.31 *Push-button-operated door with speed control*

The symbol given in Figure 7.29 is for a throttle valve which throttles fluid passing in either direction through the valve. Figure 7.32 shows the form that can be taken by a throttling valve which only throttles the flow in one direction through the valve. The fluid flow through the valve has two possible paths: through the restricted portion of the throttle valve or through the bypass of the non-return valve. When fluid flows into port A (Figure 7.32(a)), the direction of the flow is such that it forces the non-return plug to close the path through that valve and only gives a flow path through the throttle valve, i.e. the portion controlled by the movement of the screw. When fluid flows into port B (Figure 7.32(b)), the direction of flow is such that it forces the non-return plug back against its spring and so there are now two flow paths through the valve, through the throttle valve and through the bypass. Most of the fluid flows through the bypass. Thus the result is a unidirectional flow control valve, since effectively the flow is only throttled for flow in one direction through it. Figure 7.32(c) shows the circuit symbol used for the valve.

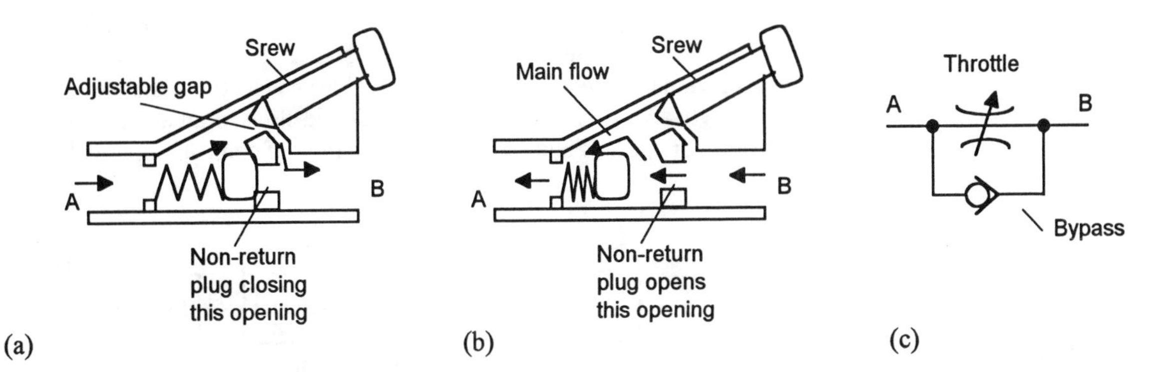

Figure 7.32 *Flow regulator with flow direction such that it has (a) non-return valve closed, (b) non-return valve open, and (c) symbol used in circuit diagrams*

To illustrate the use of such a flow regulator, Figure 7.33 shows one used for speed control with a double-acting cylinder. When the push-button 1 is pressed, the air from the supply flows through ports P and A and so into one end of the cylinder. The air that is the other side of the piston has to escape through the flow regulator. The flow regulator is, however, connected in such a way that the flow has to pass through the restriction in the valve and so cannot escape quickly. Thus the pressure on the right hand side of the piston cannot drop quickly and so slows down the movement of the piston. As a result, when push-button 1 is pressed the piston in the cylinder moves relatively slowly to its new position, the speed being determined by the adjustment of the screw in the flow regulator. When push-button 2 is pressed, the air from the supply is connected through ports P and A to the right hand side of the cylinder via the flow regulator. But this time the flow regulator is connected so that the bypass opens and so there is little restriction on the air flow. Since the air can escape rapidly from the other side of the piston, there is rapid movement of the piston to the left. The flow regulator in this case thus slows down the movement of the piston in one direction but has virtually no effect on its movement in the other direction.

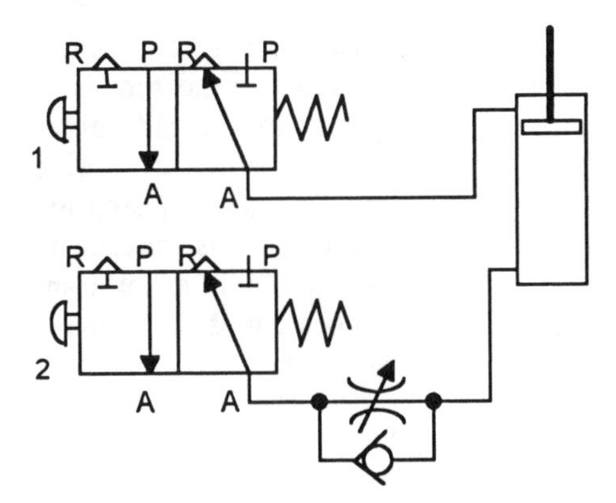

Figure 7.33 *An example illustrating the use of a flow regulator*

7.6 Servovalve

The term *servomechanism* is used for feedback control systems in which the controlled variable is a position or speed. Electro-hydraulic servo-systems generally have a controller which gives an electric signal output and uses a servovalve to control the hydraulic fluid flow to an actuator and so its position or speed. Figure 7.34 shows the type of control loop involved.

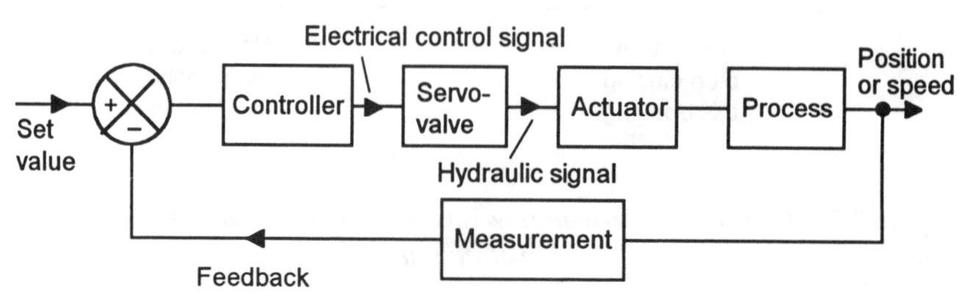

Figure 7.34 *A servo-system*

Electro-hydraulic servovalves have an input of an electrical signal which is used by a torque motor to position a spool in a valve with the result that the output is an hydraulic fluid flow to control the movement of an actuator. Figure 7.35(a) shows an example of such a valve and torque motor.

The *torque motor* essentially consists of two coils and a pivoted armature. When the input current flows through coils, a magnetic field is produced which acts on the armature and causes it to swing off-centre about its pivot. The deflection of the end of the armature is reasonably proportional to the size of the current input. The result of the movement of the end of the armature is that the spool moves in the servovalve. This movement results in the flow of hydraulic fluid from port P to the A or B ports, and from the A or B ports to the return port T, being controlled to give a flow rate which is proportional to the movement of the spool. Figure 7.35(b) shows the symbol that can be used to represent the arrangement.

The servovalve may then be used to control an actuator of a double acting cylinder and hence the position of some load, perhaps a machine tool and so produce a controlled machining operation. Figure 7.35(c) shows the form such a control system might have. The input requirement for the position of the tool is to the controller which produces an error signal which, after amplification or perhaps some other signal processing, operates the torque motor to drive the shuttle in the servovalve. This results in the movement of a piston in a double acting cylinder and hence movement of the tool. This movement is monitored by some sensor, perhaps a linear variable differential transformer (LVDT), and the output, after signal processing, is fed to the comparator where it is compared with the required set value.

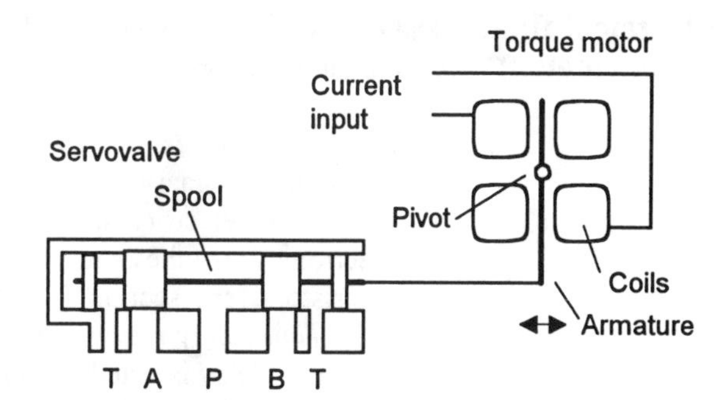

(a)

A B A B A B
P T P T P T
Solenoid operated
Solenoid operated

(b)

Signal processing
Sensor
Detects position of load
Load
Double acting cylinder
A
B
Error
Amplifier
Servovalve
Set value for position
Torque motor
T
P
Pressure relief valve
Return to reservoir
Controller
Return to reservoir
Pump
Motor
Reservoir

(c)

Figure 7.35 *(a) Servovalve with torque motor, (b) symbol, (c) example of use in a servosystem*

7.7 Electrical final control elements

The following are some of the electrical elements commonly used as correction/final control elements:

1 *Switching devices*
The control signal is applied to some form of switch which switches on or off some electrical device, perhaps a heater or a motor. Examples are electronic switches, involving such components as thyristors and transistors (see section 7.8), and relays (see section 6.2.1 and Figure 6.3). *Time-delay relays* are relays that have a delayed switching action. The time delay is usually adjustable and can be initiated when a current flows through the relay coil or when it ceases to flow through the coil.

2 *Solenoid type devices*
A current through a solenoid is used to actuate a soft iron core, as for example the solenoid-operated hydraulic/pneumatic valve where a control current through a solenoid is used to control a hydraulic/ pneumatic flow. Figure 7.36 shows the basic principle. When a current passes through the solenoid a magnetic field is produced which exerts a force on the soft iron plunger, pulling it towards the coil. Thus changing the current changes the force acting on the plunger and so its movement. This movement may be magnified by a system of levers. *Solenoid valves* are operated in a similar manner, Figure 7.37 showing the form of a solenoid operated four-way spool valve.

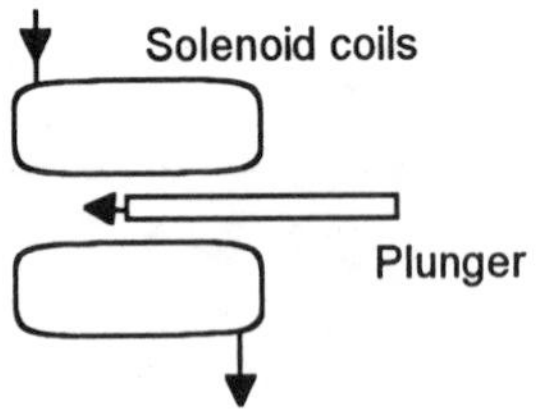

Figure 7.36 *Solenoid actuator*

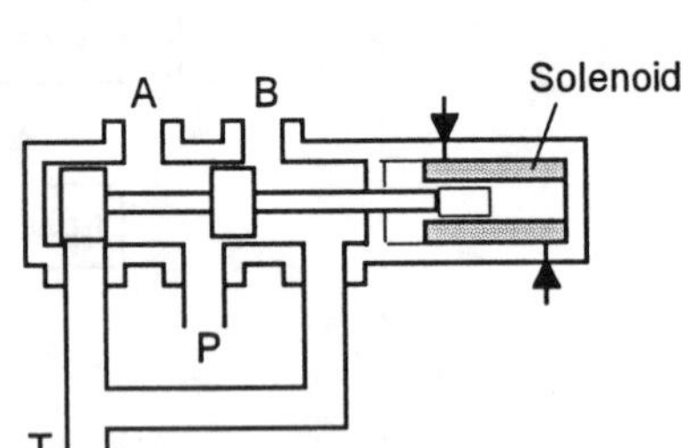

Figure 7.37 *Solenoid operated four-way spool valve*

3 *Motors*
With motors, d.c. or a.c., a current through a motor coil is used to produce rotation. See section 7.9 for more details. They might be used to rotate the arm of a robot, position a workpiece for a machine tool, etc.

7.8 Electronic switches

A *diode*, Figure 7.38 showing the circuit symbol, only allows a significant current in one direction through it. When the applied voltage is such as to give a current in the forward direction, the so-termed *forward bias* condition, the resistance of the diode is quite small; when the applied voltage gives a current in the reverse direction, the so-termed *reverse bias* condition, the resistance is high. Thus a diode can be regarded as an electronic switch which is only switched on when there is forward bias.

Figure 7.38 *Diode symbol*

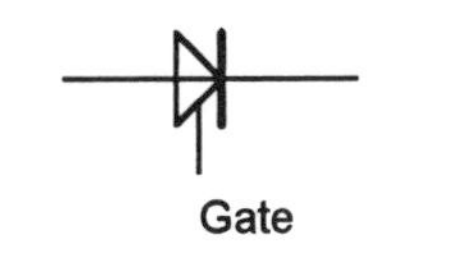

Figure 7.39 *Thyristor symbol*

The *thyristor*, Figure 7.39 showing the circuit symbol, can be regarded as a diode which has a connection, called the *gate*, controlling the conditions under which the diode can be switched on. As with the diode, the thyristor passes negligible current when reverse biased. With the gate current zero, when forward biased the current is also negligible until a forward breakdown voltage is exceeded. This might, for example, be 300 V. When this occurs the voltage across the diode falls to a low level, about 1 to 2 V, and the current is then only limited by the external resistance in a circuit. If the thyristor is in series with a resistance of, say, 20 Ω then before breakdown we have a very high resistance in series with the 20 Ω and so virtually all the 300 V is across the thyristor and there is negligible current. When forward breakdown occurs, the voltage across the thyristor drops to 2 V and so there is now 300 – 2 = 298 V across the 20 Ω resistor, hence the current rises to 298/20 = 14.9 A. When once switched on the thyristor remains on until the forward current is reduced to below a level of a few milliamps. The voltage at which forward breakdown occurs is determined by the current entering the gate, the higher the current the lower the breakdown voltage. The gate current thus determines the voltage at which the thyristor switches from a high resistance to a low resistance.

To illustrate how a thyristor can be used for control purposes, Figure 7.40 shows how a thyristor could be used to control a d.c. voltage. The thyristor is operated as a switch by using the gate to switch the device on or off. When an alternating current is supplied to the gate, the gate current rises to a value which switches on for the value of the supply voltage, then it drops to a value which switches it off, then it rises again to a value which switches it on, then it falls to a value which switches it off, and so on. The supply voltage is thus periodically switched on and off. As a result of this the average value of the output d.c. voltage can be varied and hence controlled. Thyristor control is used in this way to control the speed of motors by varying the current supplied to the motor coil.

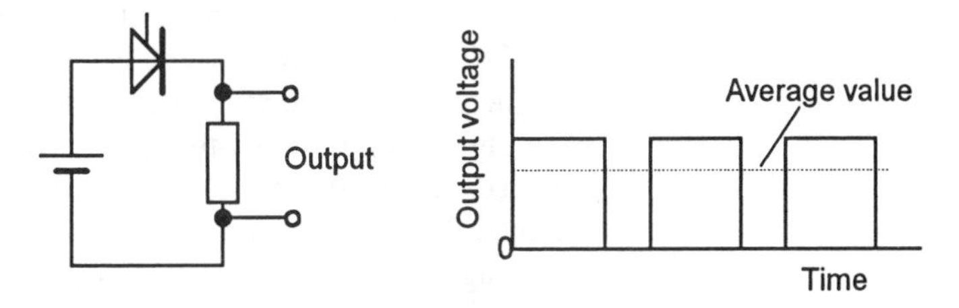

Figure 7.40 *Control of a d.c. voltage*

Bipolar transistors can be used as switches. When, for the circuit shown in Figure 7.41, there is no input voltage then virtually the entire V voltage appears at the output. When the input voltage is made sufficiently high the transistor switches so that very little of the V voltage appears at the output. Thus by changing the voltage applied to the transistor, another voltage can be switched from a high to a low value.

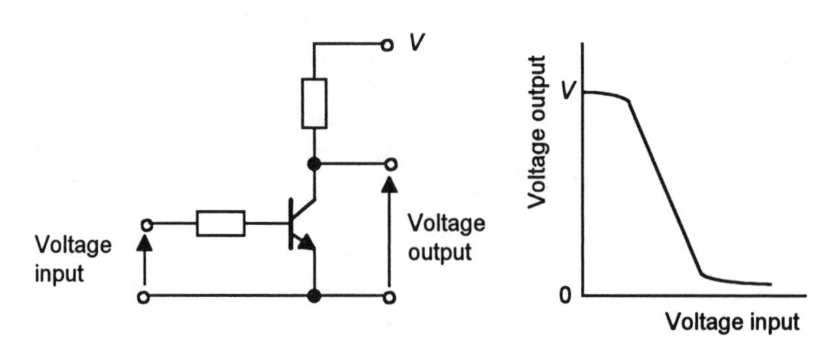

Figure 7.41 *A transistor switch circuit*

7.9 Motors

Electric motors are frequently used as the final control element in position or speed-control systems. The term *servomotor* is used for the specially designed versions of motors that are used for such control tasks. The basic principle on which motors are based is that a force is exerted on a conductor in a magnetic field when a current passes through it. For a conductor of length L carrying a current I in a magnetic field of flux density B at right angles to the conductor, the force F equals BIL.

There are many different types of motor. In the following, discussion is restricted to those types of motor that might be used as servomotors, this including the stepper motor. A *stepper motor* is a form of servomotor that is used to give a fixed and consistent angular movement by rotating an object through a specified number of revolutions or fraction of a revolution.

7.9.1 D.C. motors

In the d.c. motor, coils of wire are mounted in slots on a cylinder of magnetic material called the *armature*. The armature is mounted on bearings and is free to rotate. It is mounted in the magnetic field produced by *field poles*. This magnetic field might be produced by permanent magnets or an electromagnet with its magnetism produced by a current passing through the, so-termed, *field coils*. Whether permanent magnet or electromagnet, these generally form the outer casing of the motor and are termed the *stator*. Figure 7.42 shows the basic elements of d.c. motor with the magnetic field of the stator being produced by a current through coils of wire. In practice there will be more than one armature coil and more than one set of stator poles. The ends of the armature coil are connected to adjacent segments of a segmented ring called the *commutator* which rotates with the armature. Brushes in fixed positions make contact with the rotating commutator contacts. They carry direct current to the armature coil. As the armature rotates, the commutator reverses the current in each coil as it moves between the field poles. This is necessary if the forces acting on the coil are to remain acting in the same direction and so the continue rotation.

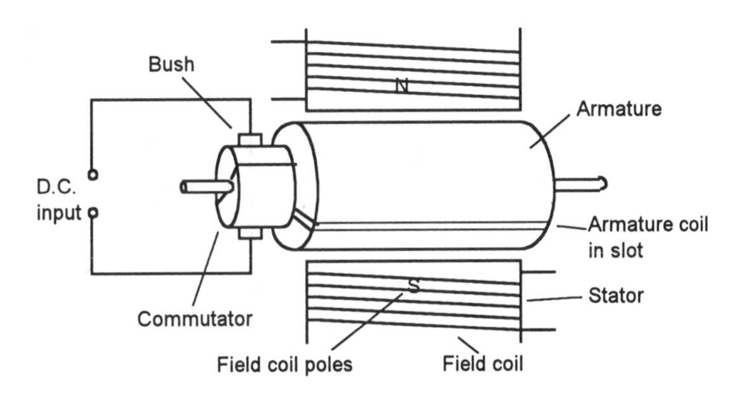

Figure 7.42 *Basic elements of a d.c. motor*

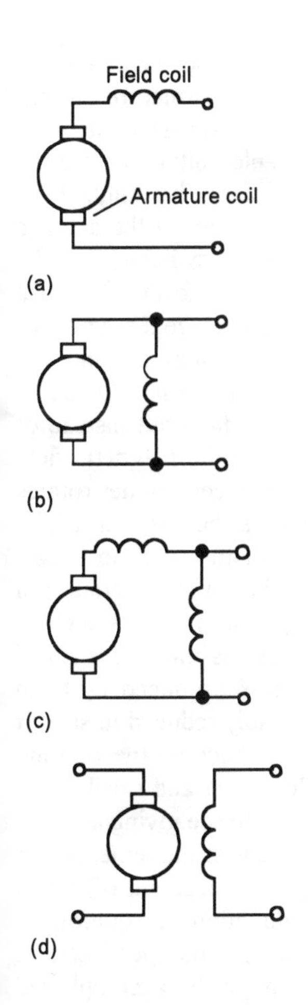

Figure 7.43 *(a) Series, (b) shunt, (c) compound, (d) separately wound d.c. motors*

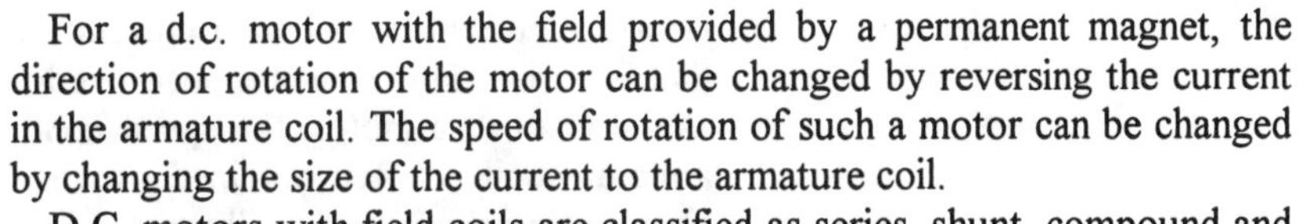

For a d.c. motor with the field provided by a permanent magnet, the direction of rotation of the motor can be changed by reversing the current in the armature coil. The speed of rotation of such a motor can be changed by changing the size of the current to the armature coil.

D.C. motors with field coils are classified as series, shunt, compound and separately excited according to how the field windings and armature windings are connected. With the *series-wound motor* the armature and fields coils are in series (Figure 7.43(a)). Such a motor exerts the highest starting torque and has the greatest no-load speed. However, with light loads there is a danger that a series-wound motor might run at too high a speed. Reversing the polarity of the supply to the coils has no effect on the direction of rotation of the motor, since both the current in the armature and the field coils are reversed. With the *shunt-wound motor* (Figure 7.43(b)) the armature and field coils are in parallel. It provides the lowest starting torque, a much lower no-load speed and has good speed regulation. It gives almost constant speed regardless of load and thus shunt wound motors are very widely used. To reverse the direction of rotation, either the armature or field current can be reversed. The *compound motor* (Figure 7.41(c)) has two field windings, one in series with the armature and one in parallel. Compound-wound motors aim to get the best features of the series and shunt-wound motors, namely a high starting torque and good speed regulation. The *separately excited motor* (Figure 7.43(d)) has separate control of the armature and field currents. The direction of rotation of the motor can be obtained by reversing either the armature or the field current. Figure 7.44 indicates the general form of the torque–speed characteristics of the above motors. The separately excited motor has a torque–speed characteristic similar to the shunt wound motor.

The choice of d.c. motor will depend on what it is to be used for. Thus, for example, with a robot manipulator the robot wrist might use a series-wound motor because the speed decreases as the load increases. A shunt-wound motor might be used if a constant speed was required, regardless of the load.

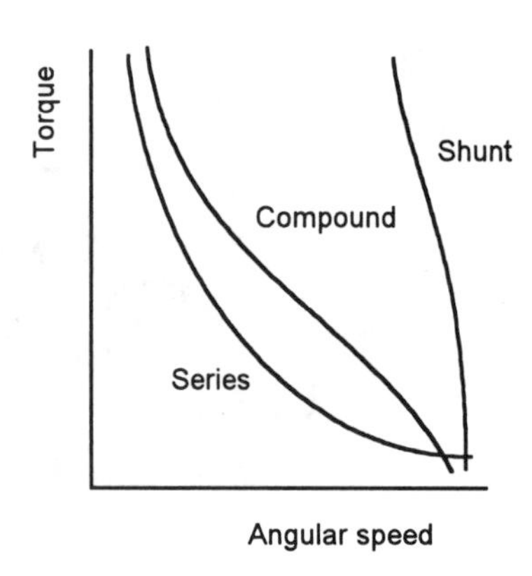

Figure 7.44 *Torque-speed characteristics of d.c. motors*

The speed of such d.c. motors can be changed by either changing the armature current or the field current. Generally it is the armature current that is varied. However, because fixed voltage supplies are generally used as the input to a motor, the speed control of a variable voltage across the armature coil is often obtained by an electronic circuit. This might be a circuit using thyristors, as illustrated in Figure 7.40, to control the average d.c. voltage by varying the time for which a d.c. voltage is switched on. The term *pulse width modulation (PWM)* is used for such a form of control since the width of the d.c. pulses is used to control the average d.c. voltage.

The above represents the general forms that can be taken by d.c. motors. A form which is particularly used for servomechanisms is the ***brushless d.c. motor***. This uses a permanent magnet for the magnetic field but instead of the arrangement being of the armature coil rotating in the magnetic field produced by a stationary permanent magnet, the permanent magnet rotates within the stationary armature coil. Figure 7.45 shows the basic principle, just one armature coil being shown. Because the armature coil no longer revolves there is no need for a commutator and brushes. The material used for the permanent magnet is one of the newer magnetic materials such as samarium cobalt. It is the development of such materials that has enabled such motors to be made since the high magnetic flux produced by them enables the magnets used with motors to be considerably reduced in size. It is this reduction in size which enables the magnet to become the rotating element. Because the rotor, i.e. the magnet, has a low mass and small size it has low inertia and so is able to quickly change speed, hence giving a motor which can follow the intricate, predetermined sequence of movements that might be required in a servomechanism controlling the movement of, say, a machine tool. The motor can be reversed in its direction of rotation by merely reversing the current through the armature coil. Its speed can be controlled by changing the armature current. This might be accomplished with a fixed voltage supply to a motor by using an electronic circuit, such as a circuit using thyristors (Figure 7.40) to control the average d.c. voltage by varying the time for which a d.c. voltage is switched on.

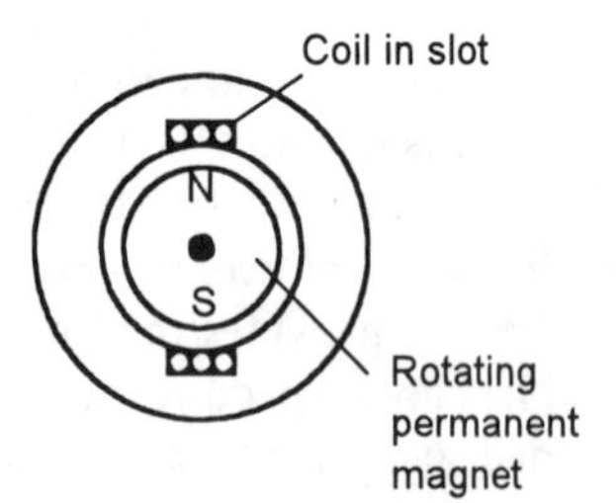

Figure 7.45 *Basic principle of the brushless d.c. motor*

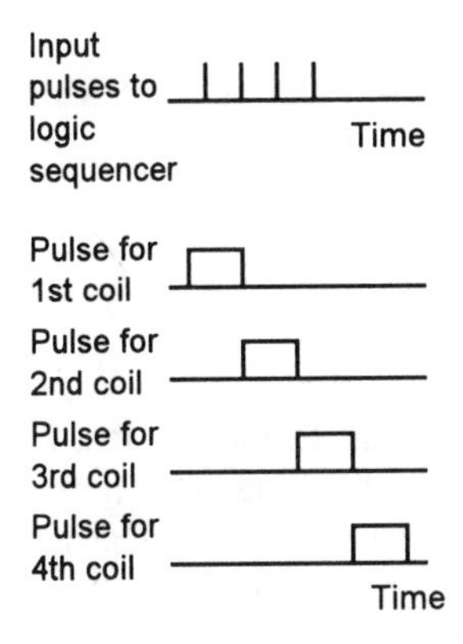

Figure 7.52 *The input and outputs of the drive system*

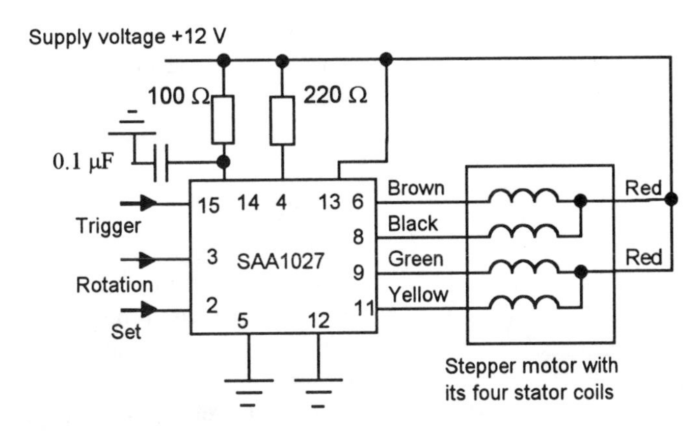

Figure 7.53 *Driver circuit involving the integrated circuit SAA1027 for a 12 V 4-phase unipolar stepper motor*

Driver circuits can be obtained as integrated circuits. For example, the driver integrated circuit SAA1027 (Figure 7.53) can be used with 4-phase unipolar stepper motors. For this circuit, the stepper motor will step once for each low to high transition on the input to pin 15, giving outputs from pins 6, 8, 9 and 11. The motor will run clockwise when pin 3 is low (<4.5 V), anti-clockwise when it is high (>7.5 V). When pin 2 is made low, the output resets. Typical values of components are included and the colour coding of the wires to a typical stepper motor. The trigger, rotation and set pulses might be supplied by a microprocessor which is programmed to deliver the pulses at the appropriate times or when signals are inputted to it.

The following are some of the terms commonly used in specifying stepping motors:

1 *Holding torque*
This is the maximum torque that can be applied to a powered motor without moving it from its rest position and causing spindle rotation.

2 *Pull-in torque*
This is the maximum torque which must be overcome for a motor will start, for a given pulse rate, and stop without losing a step.

3 *Pull-out torque*
This is the maximum torque that can be applied to a motor, running at a given stepping rate, without it failing to keep in step.

4 *Pull-in rate*
This is the maximum switching rate, i.e. the number of steps per second, at which a loaded motor can start without losing a step.

5 *Pull-out rate*
This is the switching rate, i.e. the number of steps per second, at which a loaded motor will remain in step as the switching rate is reduced.

6 *Slew range*
This is the range of switching rates between pull-in and pull-out within which the motor keeps in step but cannot start up or reverse.

Example

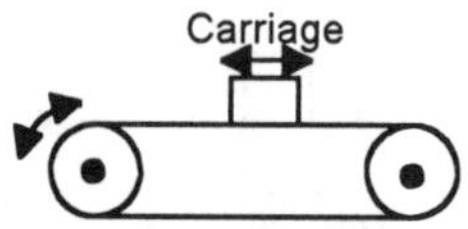

A stepper motor is to be used to drive, through a belt and pulley system (Figure 7.54), the carriage of a printer. The belt has to move a mass of 500 g which has to be brought up to a velocity of 0.2 m/s in a time of 0.1 s. Friction in the system means that movement of the carriage requires a constant force of 2 N. The pulleys have an effective diameter of 40 mm. Determine the required pull-in torque.

Figure 7.54 *Example*

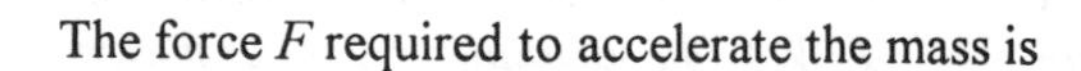

The force F required to accelerate the mass is

$$F = ma = 0.500 \times \frac{0.2}{0.1} = 1.0 \text{ N}$$

The total force that has to be overcome is the sum of the above force and that due to friction. Thus the total force that has to be overcome is $1.0 + 2 = 3$ N.

This force acts at a radius of 0.020 m and so the torque that has to be overcome to start, i.e. the pull-in torque, is

$$\text{torque} = \text{force} \times \text{radius} = 3 \times 0.020 = 0.06 \text{ N m}$$

Problems

Questions 1 to 20 have four answer options: A, B, C and D. Choose the correct answer from the answer options.

1 Decide whether each of these statements is True (T) or False (F).

For a hydraulic cylinder:
(i) The force that can be exerted by the piston is determined solely by the product of the pressure exerted on it and its cross-sectional area.
(ii) The speed with which the piston moves is determined solely by the product of the rate at which fluid enters the cylinder and the cross-sectional area of the piston.

A (i) T (ii) T
B (i) T (ii) F
C (i) F (ii) T
D (i) F (ii) F

2 A pneumatic cylinder has a piston of cross-sectional area 0.02 m^2. The force exerted by the piston when the working pressure applied to the cylinder is 2 MPa will be:

A 100 MN
B 40 MN
C 40 kN
D 20 kN

3 A hydraulic cylinder with a piston having a cross-sectional area of 0.01 m^2 is required to give a workpiece an average velocity of 20 mm/s. The rate at which hydraulic fluid should enter the cylinder is:

A 4×10^{-6} m^3/s
B 2×10^{-4} m^3/s
C 0.2 m^3/s
D [illegible] m^3/s

Questions 4 to 6 refer to Figure 7.55 which shows the symbol for a valve.

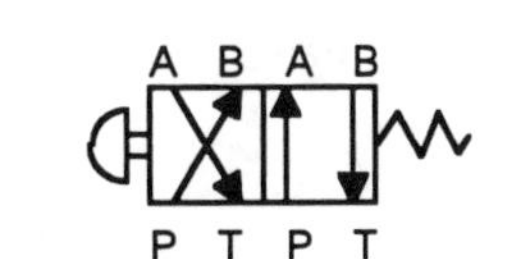

Figure 7.55 *Valve symbol*

4 Decide whether each of these statements is True (T) or False (F).

The valve has:
(i) 2 ports
(ii) 4 positions

A (i) T (ii) T
B (i) T (ii) F
C (i) F (ii) T
D (i) F (ii) F

5 Decide whether each of these statements is True (T) or False (F).

When the push button is pressed:
(i) Hydraulic fluid from the supply is transmitted through port B.
(ii) The hydraulic fluid in the line to port A is returned to the sump.

A (i) T (ii) T
B (i) T (ii) F
C (i) F (ii) T
D (i) F (ii) F

6 Decide whether each of these statements is True (T) or False (F).

When the press button is released:
(i) Hydraulic fluid from the supply is transmitted through port A.
(ii) The hydraulic fluid in the line to port B is returned to the sump.

A (i) T (ii) T
B (i) T (ii) F
C (i) F (ii) T
D (i) F (ii) F

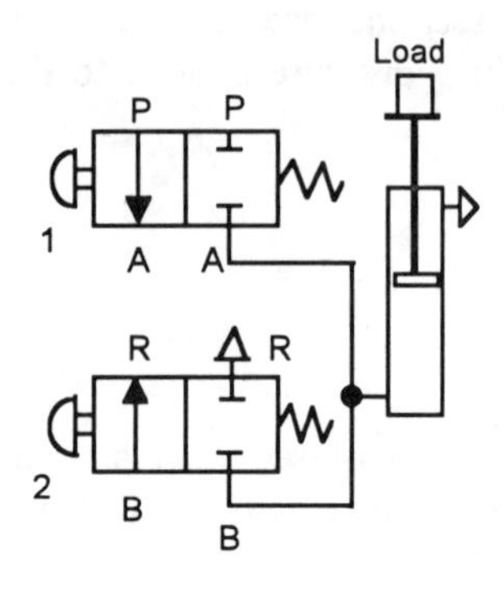

Figure 7.56 *Pneumatic circuit*

Questions 7 to 10 refer to Figure 7.56 which shows a pneumatic circuit involving two valves and a single acting cylinder.

7 Decide whether each of these statements is True (T) or False (F).

When push button 1 is pressed:
(i) The load is lifted.
(ii) Port A is closed.

A (i) T (ii) T
B (i) T (ii) F
C (i) F (ii) T
D (i) F (ii) F

8 Decide whether each of these statements is True (T) or False (F).

When push button 1, after being pressed, is released:
(i) The load descends.
(ii) Port A is closed.

A (i) T (ii) T
B (i) T (ii) F
C (i) F (ii) T
D (i) F (ii) F

9 Decide whether each of these statements is True (T) or False (F).

When push button 2 is pressed:
(i) The load is lifted.
(ii) Port B is vented to the atmosphere.

A (i) T (ii) T
B (i) T (ii) F
C (i) F (ii) T
D (i) F (ii) F

10 Decide whether each of these statements is True (T) or False (F).

When push button:
(i) 1 is pressed the load is lifted.
(ii) 2 is pressed the load descends.

A (i) T (ii) T
B (i) T (ii) F
C (i) F (ii) T
D (i) F (ii) F

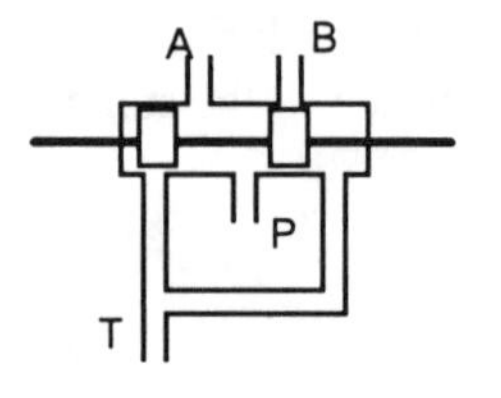

Figure 7.57 *Two-way spool valve*

11 Decide whether each of these statements is True (T) or False (F).

Figure 7.57 shows a two-way spool valve. For this valve, movement of the shuttle from left to right:
(i) Closes port A.
(ii) Connects port P to port B.

A (i) T (ii) T
B (i) T (ii) F
C (i) F (ii) T
D (i) F (ii) F

12 A flow control valve has a diaphragm actuator. The air pressure signals from the controller to give 0 to 100% correction vary from 0.02 MPa to 0.1 MPa above the atmospheric pressure. The diaphragm area needed to 100% open the control valve if a force of 400 N has to be applied to the stem to fully open the valve is:

A 0.02 m^3
B 0.016 m^3
C 0.004 m^3
D 0.005 m^3

13 Decide whether each of these statements is True (T) or False (F).

A quick-opening flow control valve has a plug shaped so that:
(i) A small change in the flow rate occurs for a large movement of the valve stem.
(ii) The change in the flow rate is proportional to the change in the displacement of the valve stem.

A (i) T (ii) T
B (i) T (ii) F
C (i) F (ii) T
D (i) F (ii) F

14 A flow control valve with a linear plug gives a minimum flow rate of 0 and a maximum flow rate of 10 m^3/s. It has a stem displacement at full travel of 20 mm and so the flow rate when the stem displacement is 5 mm is:

A 0 m^3/s
B 2.5 m^3/s
C 5.0 m^3/s
D 7.5 m^3/s

15 A flow control valve with an equal percentage plug gives a flow rate of 0.1 m^3/s when the stem displacement is 0 and 1.0 m^3/s when it is at full travel. The stem displacement at full travel is 30 mm. The flow rate with a stem displacement of 15 mm is:

A 0.32 m^3/s
B 0.45 m^3/s
C 1.41 m^3/s
D 3.16 m^3/s

16 Decide whether each of these statements is True (T) or False (F).

A flow control valve has a minimum flow rate which is 1.0% of the maximum controllable flow. Such a valve is said to have a:
(i) Rangeability of 100.
(ii) Turndown of 100.

A (i) T (ii) T
B (i) T (ii) F
C (i) F (ii) T
D (i) F (ii) F

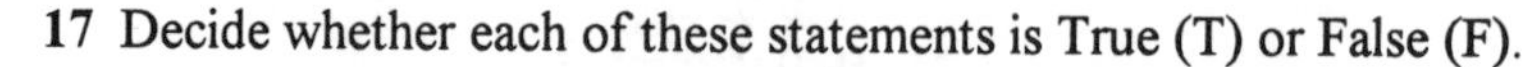

17 Decide whether each of these statements is True (T) or False (F).

Figure 7.58 shows a throttle valve in a pneumatic circuit used to control the lifting of a load by means of a single acting cylinder. The effect of the throttle valve is to:
(i) Slow down the speed at which the load is lifted.
(ii) Slow down the speed at which the load returns to the bottom after being lifted.

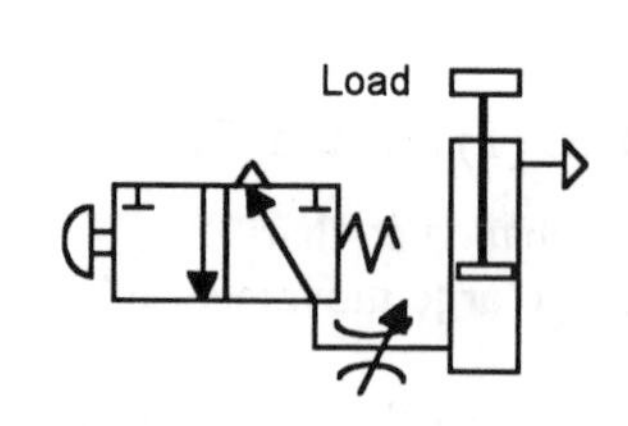

Figure 7.58 *Pneumatic circuit*

A (i) T (ii) T
B (i) T (ii) F
C (i) F (ii) T
D (i) F (ii) F

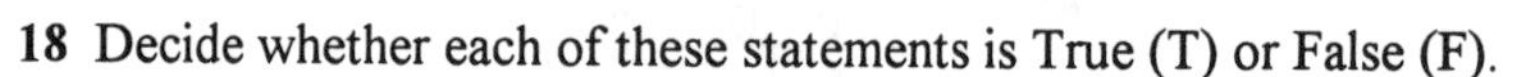

18 Decide whether each of these statements is True (T) or False (F).

A stepper motor is specified as having a step angle of 7.5°. This means that:
(i) The shaft takes 1 s to rotate through 7.5°.
(ii) Each pulse input to the motor rotates the motor shaft by 7.5°.

A (i) T (ii) T
B (i) T (ii) F
C (i) F (ii) T
D (i) F (ii) F

19 Decide whether each of these statements is True (T) or False (F).

For a series wound d.c. motor:
(i) The direction of rotation can be reversed by reversing the direction of the supplied current.
(ii) The speed of rotation of the motor can be controlled by changing the supplied current.

A (i) T (ii) T
B (i) T (ii) F
C (i) F (ii) T
D (i) F (ii) F

20 Decide whether each of these statements is True (T) or False (F).

With a shunt wound d.c. motor:
(i) The direction of rotation can be changed by reversing the direction of the armature current.
(ii) The direction of rotation can be changed by reversing the direction of the current to the field coils.

A (i) T (ii) T
B (i) T (ii) F
C (i) F (ii) T
D (i) F (ii) F

21 A force of 400 N is required to fully open a pneumatic flow control valve having a diaphragm actuator. What diaphragm area is required if the gauge pressure from the controller is 100 kPa?

22 An equal percentage flow control valve has a rangeability of 25. If the maximum flow rate is 50 m^3/s, what will be the flow rate when the valve is one-third open?

23 A stepper motor has a step angle of 7.5°. What digital input rate is required to produce a rotation of 10.5 rev/s?

24 A control valve is to be selected to control the rate of flow of water into a tank requiring a maximum flow of 0.012 m^3/s. The permissible pressure drop across the valve at maximum flow is 200 kPa. What valve size is required ? Use Table 7.1. The density of water is 1000 kg/m^3.

25 A control valve is to be selected to control the flow of steam to a process, the maximum flow rate required being 0.125 kg/s. The permissible pressure drop across the valve at maximum flow is 40 kPa. What valve size is required? Use Table 7.1. The specific volume of the steam is 0.6 m^3/s.

8 Control

8.1 Control system examples

A control system is any system that maintains a desired value of some variable or sequence of events. Systems that are used to control speed or position are termed servo-systems. This chapter is a consideration of some examples of control systems, drawing on the principles and components discussed in earlier chapters in this book.

Following the discussion of a range of examples, more detailed consideration is given in the next section to the control of position and speed, e.g. that of the movement of machine tools, and in the following section to control using programmable logic controllers (PLCs).

8.1.1 Control of fuel pressure

The modern car involves many control systems. For example, there is the *engine management system* aimed at controlling the amount of fuel injected into each cylinder and the time at which to fire the spark for ignition. Part of such a system is concerned with delivering a constant pressure of fuel to the ignition system. Figure 8.1(a) shows the elements involved in such a system. The fuel from the fuel tank is pumped through a filter to the injectors, the pressure in the fuel line being controlled to be 2.5 bar (2.5×0.1 MPa) above the manifold pressure by a regulator valve. Figure 8.1(b) shows the principles of such a valve. It consists of a diaphragm which presses a ball plug into the flow path of the fuel. The diaphragm has the fuel pressure acting on one side of it and on the other side is the manifold pressure and a spring. If the pressure is too high, the diaphragm moves and opens up the return path to the fuel tank for the excess fuel, so adjusting the fuel pressure to bring it back to the required value.

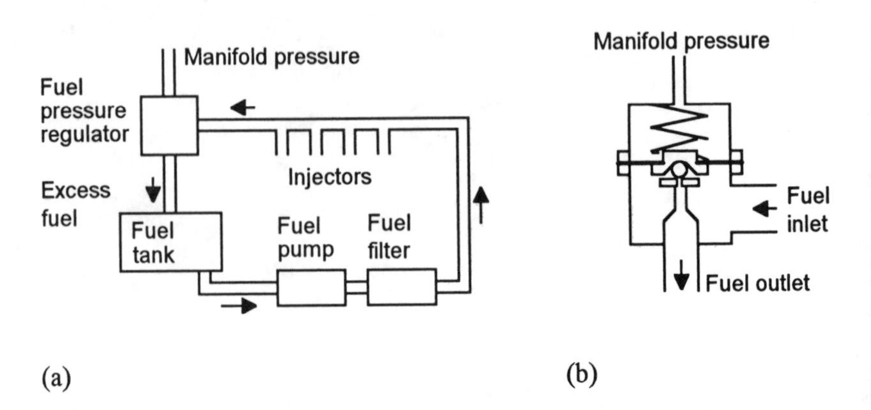

Figure 8.1 *(a) Fuel supply system, (b) fuel pressure regulator*

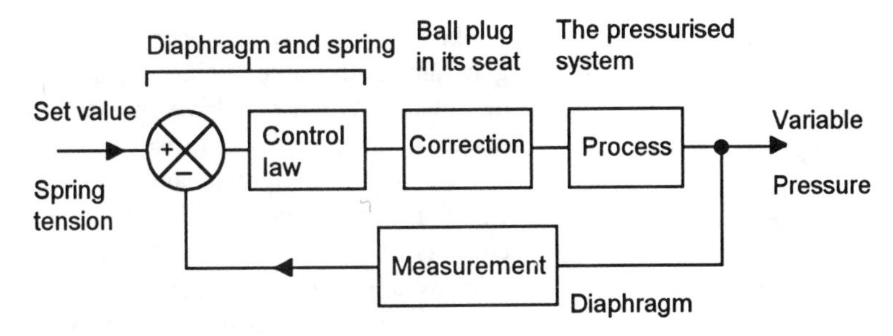

Figure 8.2 *The pressure regulator*

The pressure control system can be considered to be represented by the closed loop system shown in Figure 8.2. The set value for the pressure is determined by the spring tension. The comparator and control law is given by the diaphragm and spring. The correction element is the ball in its seating and the measurement is given by the diaphragm.

8.1.2 Antilock brake system

Another example of a control system used with a car is the *antilock brake system (ABS)*. If one of more of the vehicle's wheels lock, i.e. begins to skid, during braking, then braking distance increases, steering control is lost and tyre wear increases. Antilock brakes are designed to eliminate such locking. The system is essentially a control system which adjusts the pressure applied to the brakes so that locking does not occur. This requires continuous monitoring of the wheels and adjustments to the pressure to ensure that, under the conditions prevailing, locking does not occur. Figure 8.3 shows the principles of such a system.

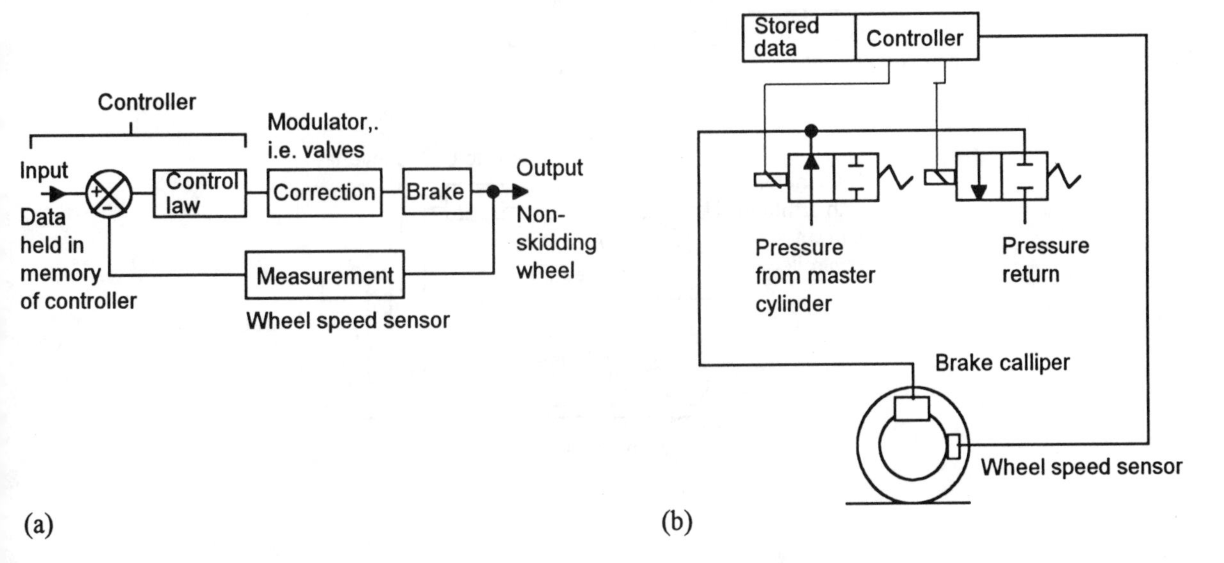

Figure 8.3 *Antilock brakes, the closed loop system in (a) block form, (b) schematic diagram*

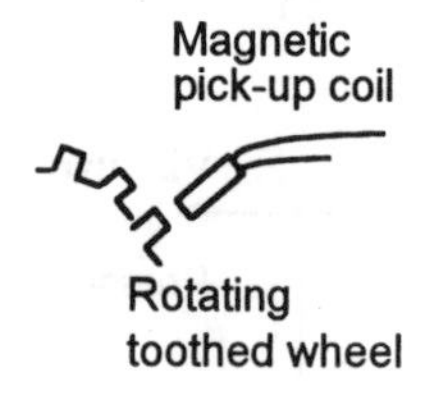

Figure 8.4 *Principle of the sensor*

The two values used to control the pressure are solenoid-operated valves, generally both valves being usually combined in a component termed the modulator. When the driver presses the brake pedal, a piston moves in a master cylinder and pressurises the hydraulic fluid. This pressure causes the brake calliper to operate and the brakes to be applied. The speed of the wheel is monitored by means of a sensor (Figure 8.4). This generally consists of a toothed wheel which rotates with the road wheel. A coil of wire, wrapped round a core of magnetic material, is near the rim of the toothed wheel and as the teeth pass the coil they change the magnetic flux in the coil and a voltage pulse is produced. The frequency of the pulses is proportional to the wheel speed. When the wheel locks, its speed changes abruptly and so the feedback signal from the sensor changes. This feedback signal is fed into the controller where it is compared with what signal might be expected on the basis of data stored in the controller memory. The controller can then supply output signals which operate the valves and so adjust the pressure applied to the brake.

8.1.3 Thickness control

As an illustration of the use of control systems in industry, Figure 8.5 shows the type of system that might be used to control the *thickness of sheet* produced by rollers.

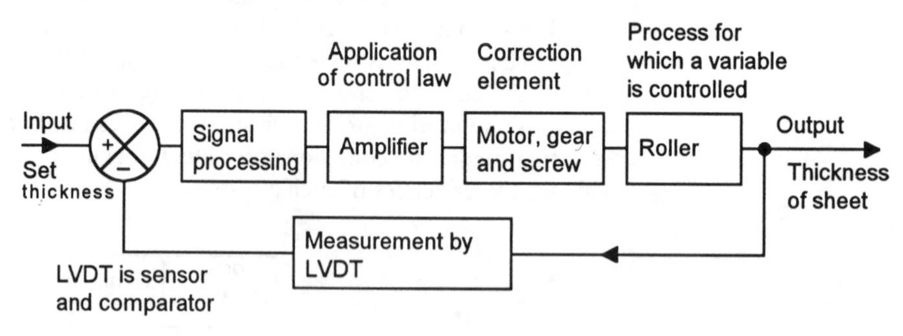

(a)

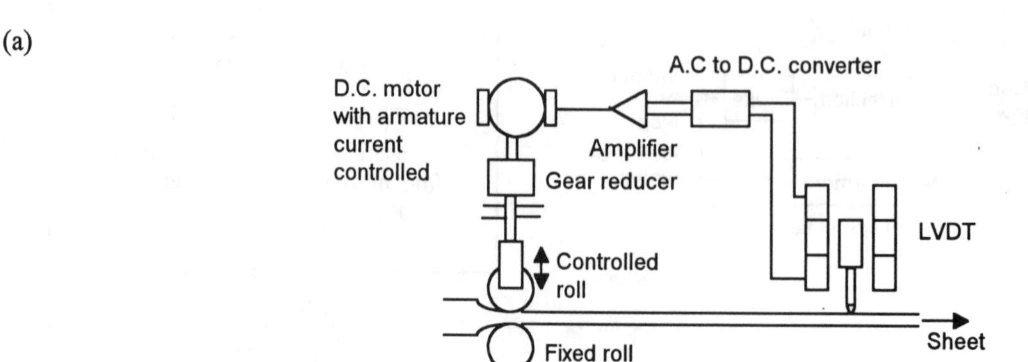

(b)

Figure 8.5 *Sheet thickness control system: (a) block diagram, (b) schematic diagram*

The thickness of the sheet is monitored by a linear variable differential transformer (LVDT). The position of the LVDT probe is set so that when the required thickness sheet is produced, there is no output from the LVDT. The LVDT produces an alternating current output, the amplitude of which is proportional to the error. This is then converted to a d.c. error signal which is fed to an amplifier. The amplified signal is then used to control the speed of a d.c. motor, generally being used to vary the armature current. The rotation of the shaft of the motor is likely to be geared down and then used to rotate a screw which alters the position of the upper roll, hence changing the thickness of the sheet produced.

8.1.4 Control of liquid level

As another example, consider a control system used to control the level of liquid in a tank using a float-operated pneumatic controller (Figure 8.6). When the level of the liquid in the tank is at the required level and the inflow and outflows are equal, then the controller valves are both closed. If there is a decrease in the outflow of liquid from the tank, the level rises and so the float rises. This causes point P to move upwards. When this happens, the valve connected to the air supply opens and the air pressure in the system increases. This causes a downward movement of the diaphragm in the flow control valve and hence a downward movement of the valve stem and the valve plug. This then results in the inflow of liquid into the tank being reduced. The increase in the air pressure in the controller chamber causes the bellows to become compressed and move that end of the linkage downwards. This eventually closes off the valve so that the flow control valve is held at the new pressure and hence the new flow rate.

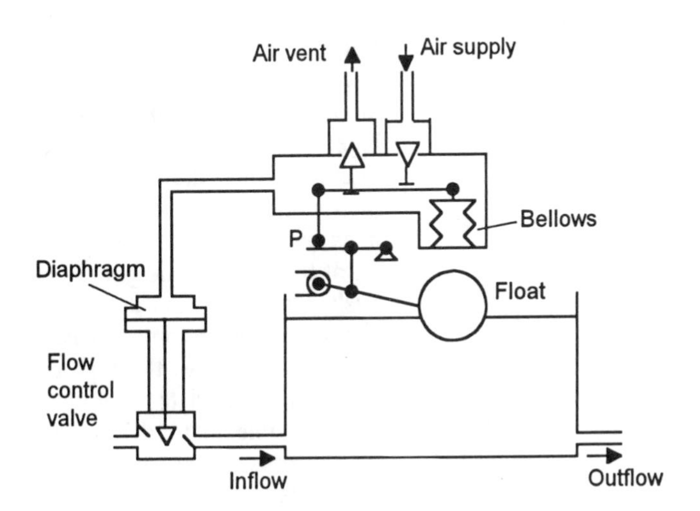

Figure 8.6 *Schematic diagram of level control system*

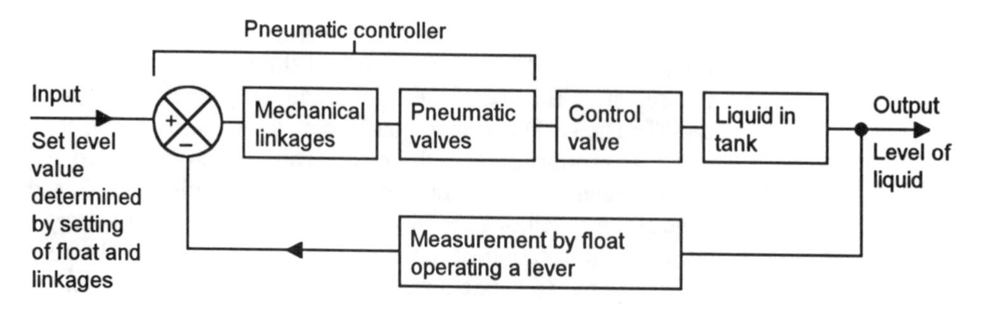

Figure 8.7 *Block diagram of level control system*

If there is an increase in the outflow of liquid from the tank, the level falls and so the float falls. This causes point P to move downwards. When this happens, the valve connected to the vent opens and the air pressure in the system decreases. This causes an upward movement of the diaphragm in the flow control valve and hence an upward movement of the valve stem and the valve plug. This then results in the inflow of liquid into the tank being increased. The bellows react to this new air pressure by moving its end of the linkage, eventually closing off the exhaust and so holding the air pressure at the new value and so the flow control valve at its new flow rate setting. Figure 8.7 shows the system represented as a block diagram.

8.1.5 Robot control systems

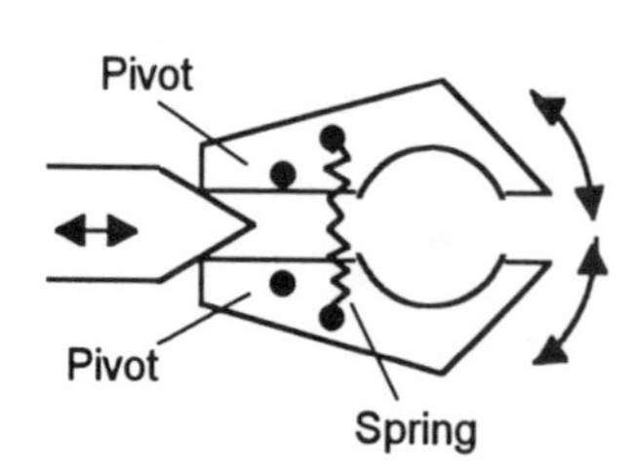

Figure 8.8 *An example of a gripper*

The term *robot* is used for a machine which is a reprogrammable multi-function manipulator designed to move tools, parts, materials, etc. through variable programmed motions in order to carry out specified tasks. Here just one aspect will be considered, the gripper used by a robot at the end of its arm to grip objects. A common form of gripper is a device which has 'fingers' or 'jaws'. The gripping action then involves these clamping on the object. Figure 8.8 shows one form such a gripper can take if two gripper fingers are to close on a parallel sided object. When the input rod to the fingers moves towards the fingers pivot about their pivots and move closer together. When the rod moves outwards, the fingers move further apart. Such motion needs to be controlled so that the grip exerted by the fingers on an object is just sufficient to grip it, too little grip and the object will fall out of the grasp of the gripper and too great might result in the object being crushed or otherwise deformed. Thus there needs to be feedback of the forces involved at contact between the gripper and the object.

Figure 8.9 shows the type of closed-loop control system involved. The drive system used to operate the gripper can be electrical, pneumatic or hydraulic. Pneumatic drives are very widely used for grippers because they are cheap to install, the system is easily maintained and the air supply is easily linked to the gripper. Where larger loads are involved, hydraulic drives can be used. See section 7.6 for a possible form of pneumatic/ hydraulic drive.

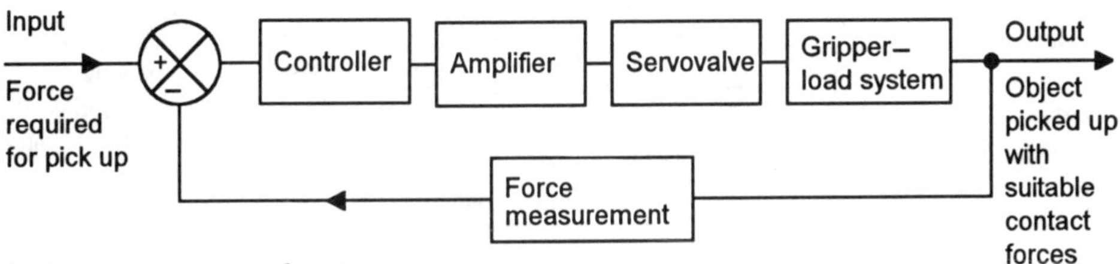

Figure 8.9 *Robot gripper control system*

Sensors that might be used for measurement of the forces involved are piezoelectric sensors or strain gauges. When a wafer of a piezoelectric material is stretched or compressed as a result of forces, the deformation results in a small e.m.f. being generated. This can be amplified and fed to the robot controller. Piezoelectric sensors when arranged in the form of a matrix can be used to sense the surface contour of the object. Strain gauges are essentially flat lengths of wire or foil which can be stuck to surfaces. They change resistance when subject to strain. Thus when strain gauges are stuck to the surface of the gripper and forces applied to a gripper, the strain gauges will be subject to strain and give a resistance change. This resistance change is thus related to the forces experienced by the gripper when in contact with the object being picked up.

The robot arm with gripper is also likely to have further control loops to indicate when it is in the right position to grip an object. Thus the gripper might have a control loop to indicate when it is in contact with the object being picked up; the gripper can then be actuated and the force control system can come into operation to control the grasp. The sensor used for such a control loop might be a microswitch which is actuated by a lever, roller or probe coming into contact with the object.

8.1.6 Machine tool control

Machine tool control systems are used to control the position of a tool or workpiece and the operation of the tool during a machining operation. The term *numerical control* is used when the information for controlling the machine is in the form of numbers representing such features as the dimensions of the workpiece, feed rate, spindle speed, etc.

Closed loop systems involve the continuous monitoring of the movement and position of the work tables on which tools are mounted while the workpiece is being machined (Figure 8.10). The amount and direction of movement required in order to produce the required size and form of workpiece is the input to the system, this being a program of instructions fed into a memory which then supplies the information as required. The term *part program* is used for this program containing all the information required for the machining of a component. The sequence of steps involved is then:

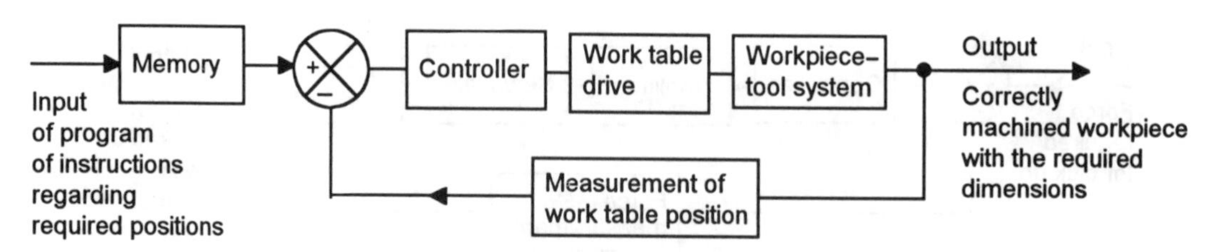

Figure 8.10 *Closed loop machine tool control system*

1. An input signal is fed from the memory store.
2. The error between this input and the actual movement and position of the work table is the error signal which is used to apply the correction. This may be an electric motor to control the movement of the work table. The work table then moves to reduce the error so that the actual position equals the required position.
3. The next input signal is fed from the memory store.
4. Step 2 is then repeated.
5. The next input signal is fed from the memory store.

And so on.

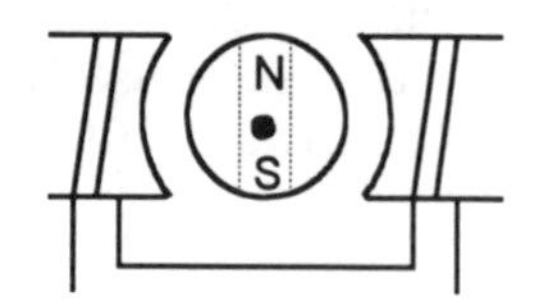

Figure 8.11 *Principle of the tacho-generator*

A separate feedback loop is required for each feature of the machine tool which is controlled. Thus each linear or rotary movement has to be monitored by some sensor. If the tool spindle speed is to be controlled then a sensor is also needed for that. A sensor that is widely used is the tacho-generator. This is essentially an alternating current generator. One form consists of a permanent magnet rotor which rotates and generates an e.m.f. in fixed stator coils (Figure 8.11), the size of the e.m.f. being a measure of the angular speed of the rotor. The linear motion of the work table can be monitored by mounting the tachogenerator on the end of the leadscrew that moves the table and so monitoring its linear movement by the rotation of the screw.

An open-loop machine tool control system (Figure 8.12) typically has a part program of instructions fed, via an input unit, to a stepper motor. This is used to control the movement of the work table. There is no feedback of the work table position. The system relies on the accuracy with which the stepper motor can set the position of the work table.

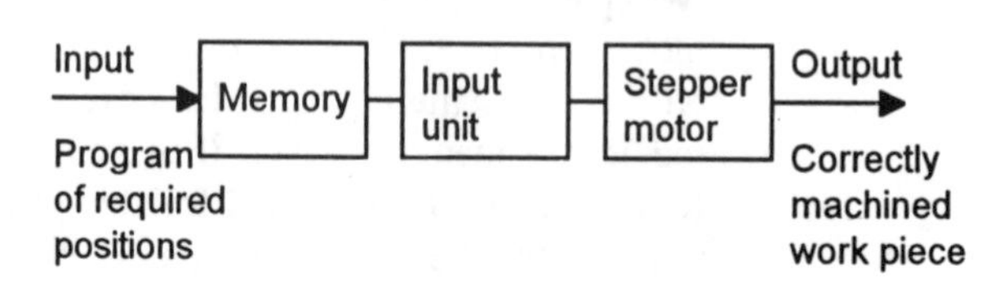

Figure 8.12 *Open-loop machine tool control system*

8.1.7 Positional control with pen plotters

As another example of a control system, consider a pen plotter which takes the output from a computer and plots a drawing on a sheet of paper. The plotter head has to accurately follow the input signal as it varies and as the sheet of paper proceeds at a controlled rate past the plotter head. What is thus required for the plotter head is an accurate control system for its displacement. Figure 8.13 shows in a block diagram the form such a control system might take.

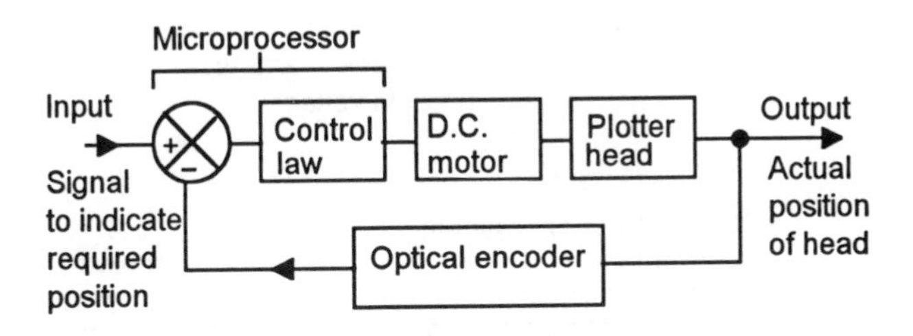

Figure 8.13 *The control system for the plotter head*

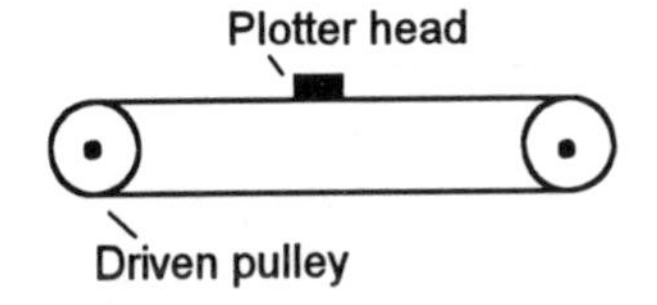

Figure 8.14 *Plotter head drive system*

A d.c. motor is used as the actuator to drive by means of a belt the movement of the plotter head (Figure 8.14). The feedback signal is provided by an optical encoder (see section 2.2.1). Typically this might give 2000 encoder counts per revolution of the motor shaft and with a pulley wheel of, say, diameter 20 mm this will mean that the position of the plotter head can be set to about $(\pi \times 20)/2000 = 0.03$ mm. The controller is likely to be a microprocessor.

The feedback from the optical encoder indicates the actual position of the plotter head. This is compared by the comparator with the signal from the computer indicating the required position. The error signal is then used to drive the motor and so move the plotter head to reduce the error to zero. As the input signal from the computer changes, so the error changes and the head has to move to maintain zero error.

8.2 Control of position and speed

Figure 8.15 shows a simple control system for the control of speed and hence displacement of a load such as a machine tool. The input is a voltage which is set by the displacement of the slider of a potentiometer. This is compared with the feedback voltage by an operational amplifier and the output used, via a servo amplifier, to control the armature current of a d.c. motor. A servo amplifier is one that allows for polarity reversal in order that the motor can run in either direction. The rotation of the motor shaft is transformed into the required linear motion by using a step-down gear to rotate a screw and so give a linear displacement of the load. The load could be a work piece being moved for machining. The position of the load is fed back to the comparator by using the rotation of the screw to rotate the slider on a linear potentiometer and so give a voltage signal related to the position.

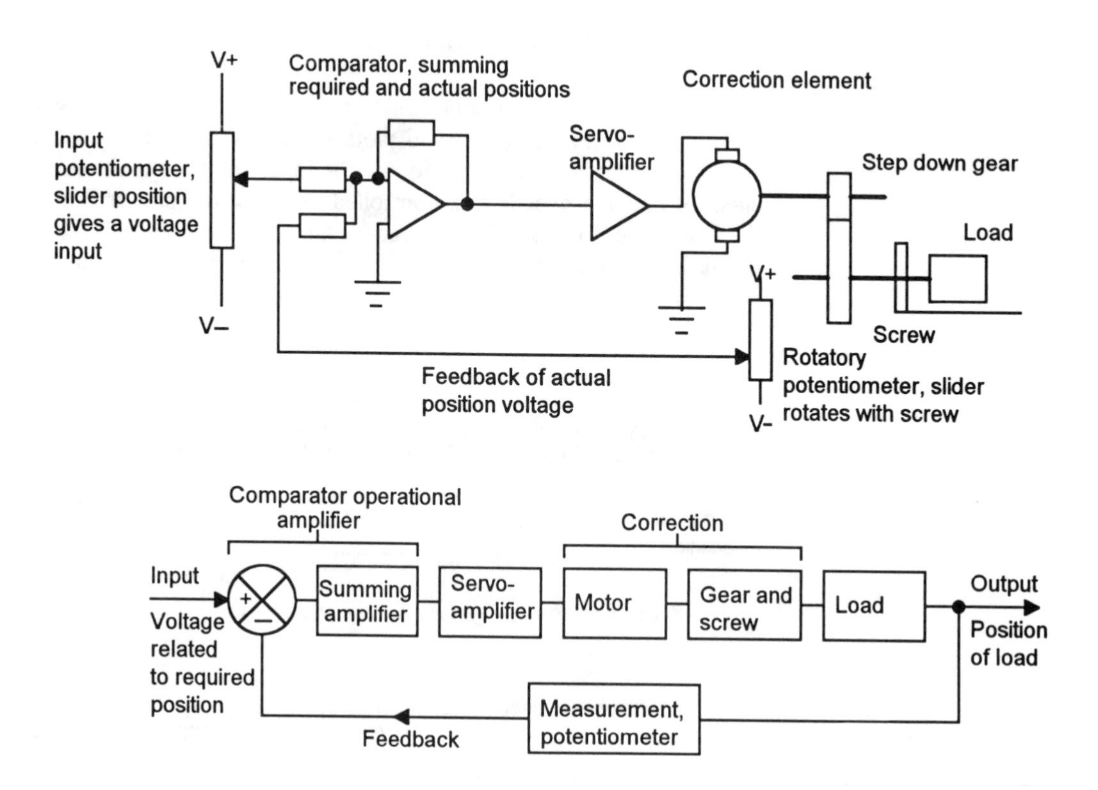

Figure 8.15 *Basic form of a position control system*

Consider what might happen with the above system if there is a sudden change in the input. The result is a sudden increase in the error signal and so a corresponding increase in current to the motor armature. The output shaft of the motor then accelerates rapidly. But it will be some time before the effect of the increased rotation of the motor shaft is communicated back via the feedback loop to the comparator. A consequence of this is that the motor shaft is likely to rotate past the required rotation before the control system is able to recognise that it has reached the required rotation. The result of this is that there is oscillation about the required rotation before the system settles down to the steady state value. The rapidity with which these oscillations die away depends on the damping in the system. If the gain of the amplifier is small, the degree of damping required is also small. A high gain, and hence a large signal to the correction element, is needed if the system is to respond quickly to changes in input. However, if the gain is large, then the degree of damping required is large. The difficulty in providing an adequate amount of damping limits the amplifier gain that can be used and so the speed of response of the system to a change in input.

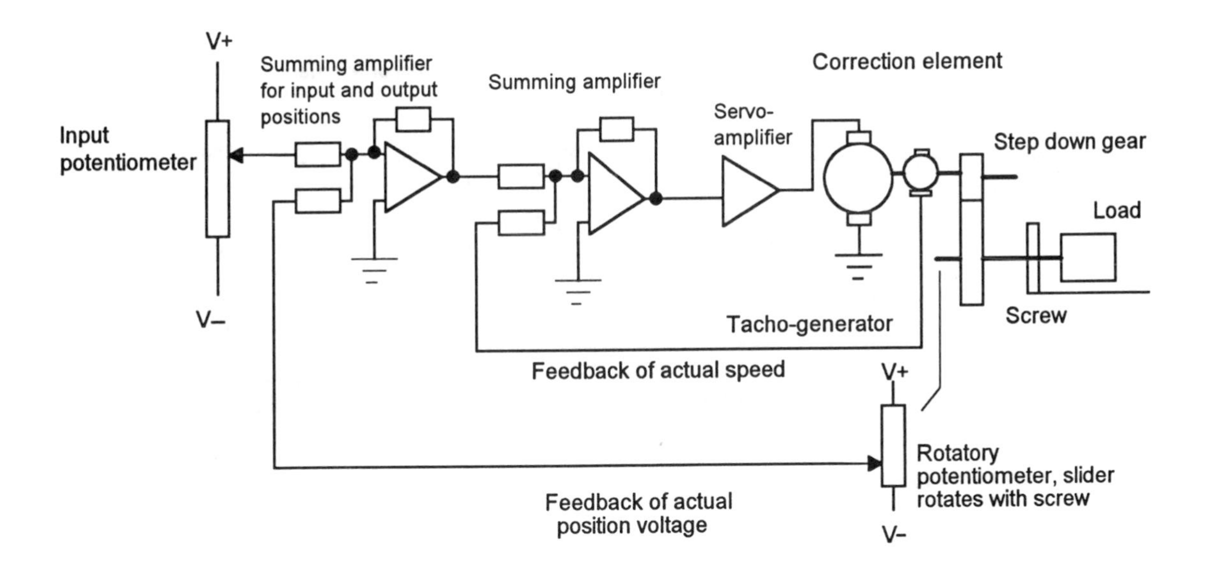

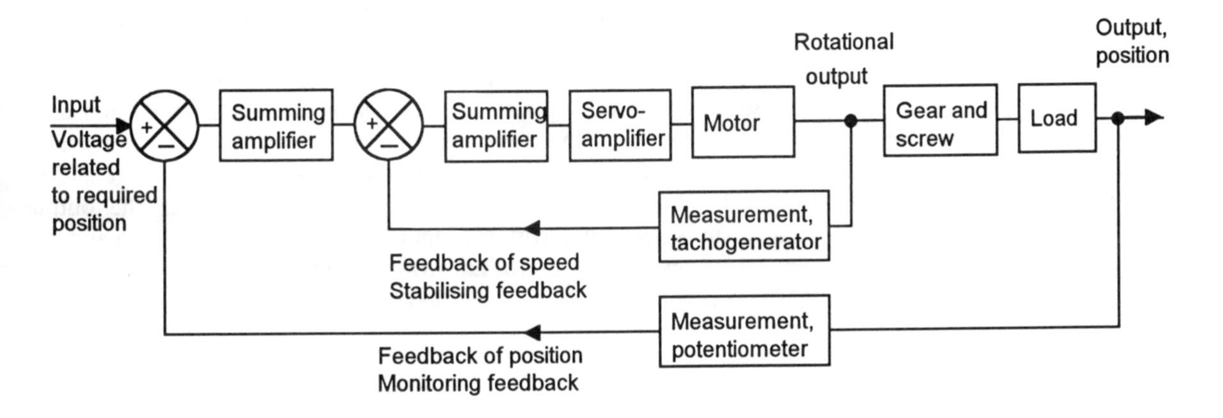

Figure 8.16 *The addition of a second feedback loop*

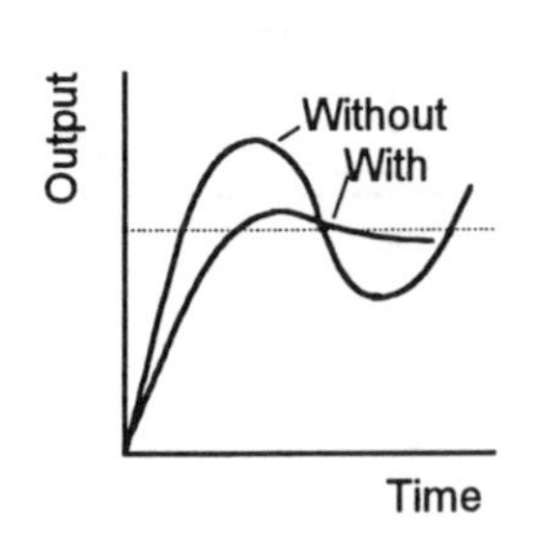

Figure 8.17 *The effect of having a tacho-generator on the response*

One way of overcoming this problem is to use a second feedback loop which gives a measurement related to the rate at which the displacement is changing. A tachogenerator is such a sensor. Thus the system used might be of the form shown in Figure 8.16. When the motor is rotating, the tachogenerator gives a voltage proportional to the speed. This is fed back and subtracted from the error voltage given by the summing amplifier used to compare the input and output positions. The tachogenerator is in fact giving a signal which is proportional to the rate of change of the displacement. Thus when there is a sudden change in the input to the control system, with the result that the motor shaft has a sudden acceleration, the tachogenerator gives an output which damps down the rate of change (Figure 8.17). The result is less oscillation. The tachogenerator has thus provided a stabilising feedback loop.

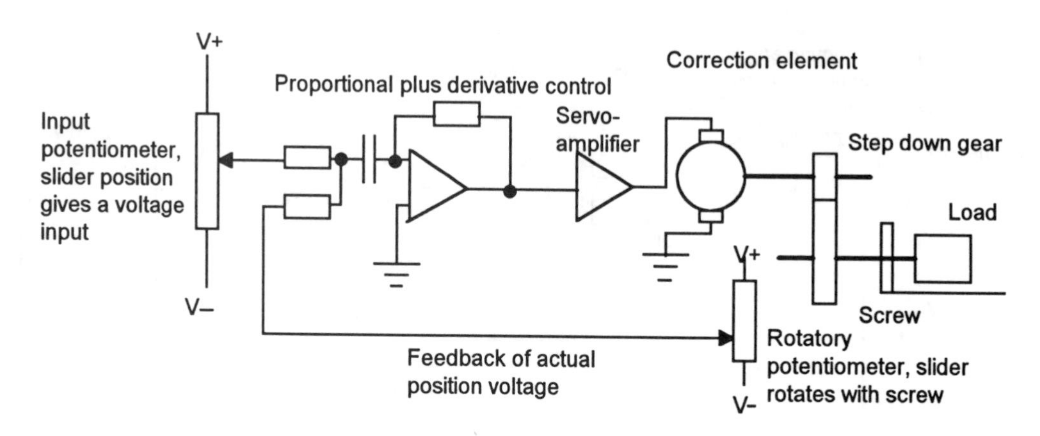

Figure 8.18 *Proportional plus derivative control*

An alternative, which can give the same result as the velocity feedback, is to employ proportional plus derivative control using operational amplifiers (Figure 8.18). The system with a tachogenerator feedback or the operational amplifier system giving proportional plus derivative control has the disadvantage of having a steady state error (see sections 6.3 and 6.4). This means that when the displacement input is changed to a new value there must be an error signal for the output displacement to change and be maintained. Thus we have to have an error signal and so the output displacement is not in step with the input displacement. This can be avoided if we add integral control to the arrangement. Figure 8.19 shows the operational amplifier circuit that would achieve this.

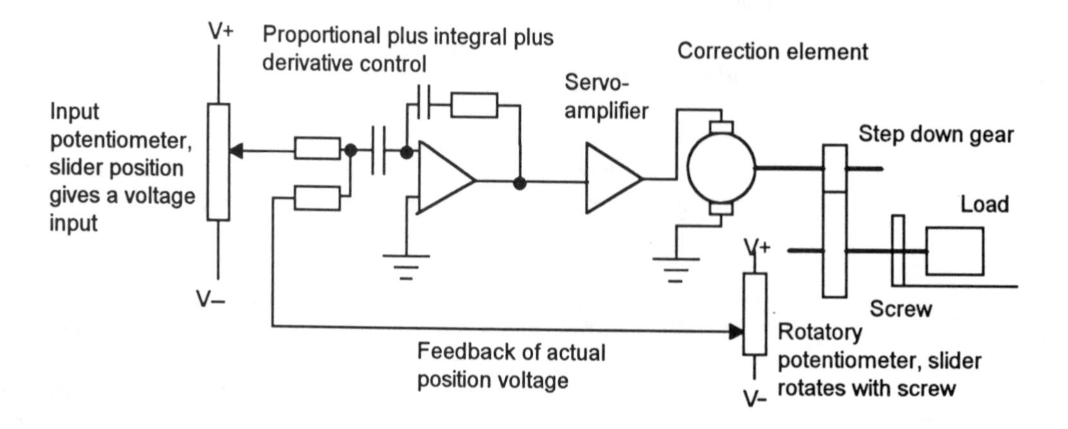

Figure 8.19 *Proportional plus integral plus derivative control*

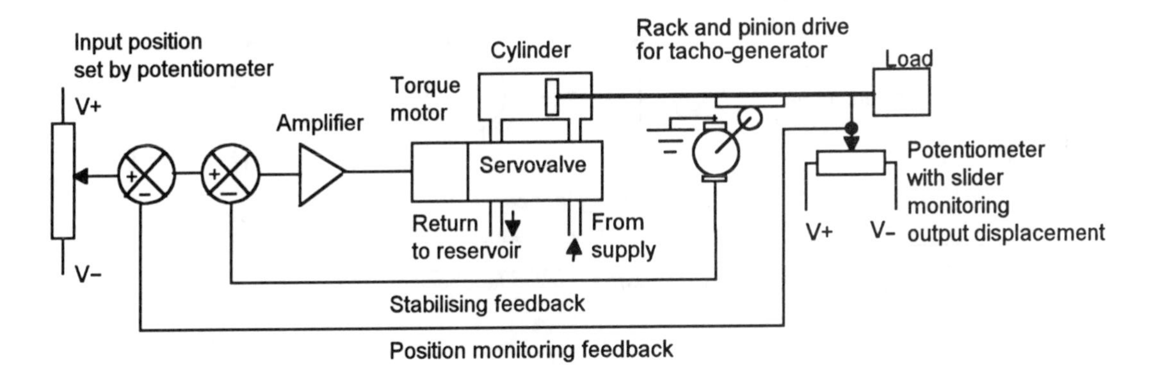

Figure 8.20 *Hydraulic servo valve used for a position control system*

The above examples have all involved the control of displacement when the final control element is an electric motor. Hydraulic and pneumatic control systems can also be used. Figure 8.20 illustrates a basic system involving an hydraulic servo valve (see section 7.6) used with electrical feedback. It shows the system with both a position error feedback loop and a tachogenerator giving a stabilising signal feedback. The comparators could be operational amplifier circuits.

The traditional method of specifying the required position input and giving feedback about the position output is a potentiometer, as in the examples given above. This method suffers from the problem that if there is any change in the voltage applied across the potentiometer then the output signal taken from the potentiometer will change, even when its position remains unchanged. Because of this, applications requiring high positional accuracy often use an optical encoder (see section 2.2.1) to give the feedback signal of the actual displacement. The input giving the required value of the position might then be via a microprocessor. With the microprocessor the input involves digital signals which may be entered using a keyboard. Slight changes in the voltages of these signals will have no effect since effectively the input is a series of on/off signals.

Figure 8.21 illustrates the form such a position control system might take. The d.c. motor has an optical encoder and tachogenerator mounted on its shaft, the tachogenerator being to give a stabilising feedback. The microprocessor computes the actual position of the motor shaft by counting the number of pulses fed back from the optical encoder and then generates an error signal by subtracting the actual position from the required position. The required position information might be inputted to the microprocessor from a keyboard or a program stored in a memory. The digital output from the microprocessor is then converted into analogue form by a digital to analogue converter. This signal then has the feedback signal from the tachogenerator combined with it. This difference is then used to drive the motor.

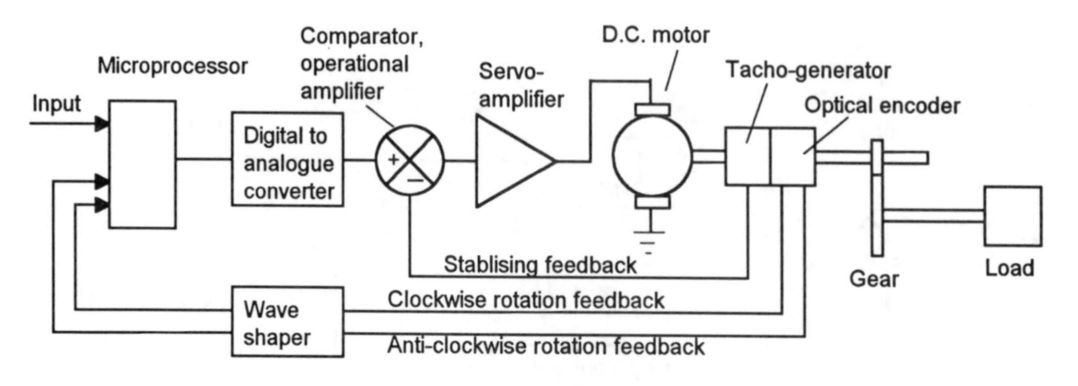

Figure 8.21 *Microprocessor controlled motor for position control*

For the motors described in the above examples, the error signal has been transmitted to the motor via a servo-amplifier. Such an amplifier is able to drive the motor in both directions. An alternative is to use pulse-width modulation (PWM) (see section 7.9.1). This involves turning a constant direct voltage on or off at a frequency determined by the error signal, thus varying the duration of pulses and so the average value of the voltage applied to the motor armature. Pulse width modulators for such an application are available as integrated circuit packages.

8.3 Control using PLCs

Programmable logic controllers (PLCs) were discussed in section 6.9. Here we consider some applications of such controllers and more of the ways they can be programmed.

Consider the situation where a PLC is to be used as an on/off controller for a heater used in temperature control. The heater is to be switched on when the actual temperature falls below the required temperature and switched off when the actual temperature is at or above the required temperature. The sensor used to monitor the actual temperature might be a thermocouple or a thermistor. It, and its associated circuitry, will give a signal which is related to the temperature. This signal, along with a voltage indicative of the required temperature, is fed to a comparator which is designed to give a high output signal when the actual temperature is above the required temperature and a low signal when it is below the required temperature. Thus when the temperature falls from above the required value to below it, the output from the comparator switches from a high to a low output. This transition can be used as the input to a PLC so that is acts and gives an on output to the heater. Figure 8.22 illustrates this for a Mitsubishi PLC.

The ladder diagram used to program the PLC for this operation is shown in Figure 8.23. The input port is taken to be X400 and the output port Y430. The switch connected to X400 has input contacts that are normally closed. When the input is low the contacts remain closed, when the input becomes high the switch opens. Thus the PLC is programmed to switch off

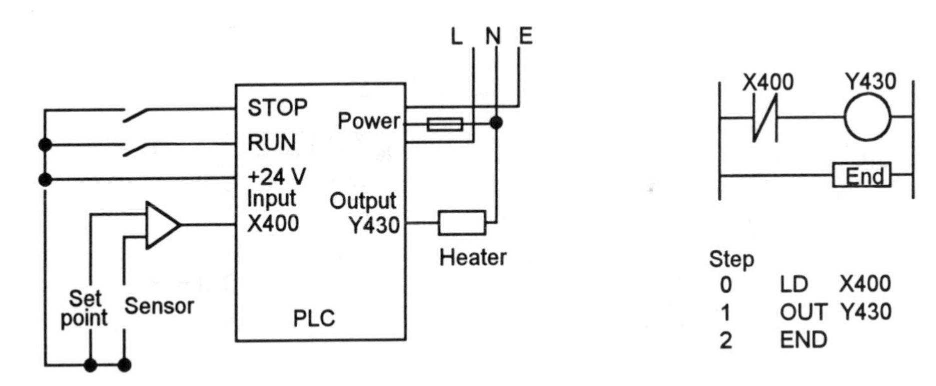

Figure 8.22 *PLC connections*

Figure 8.23 *Ladder diagram and program*

the heater when the input is high and on when the input is low. The End line is to indicate that the program consists of just the single line.

Consider now a more complex example involving the control for a central heating system (Figure 8.24). The boiler, which is thermostatically controlled, supplies hot water to the radiator system in the house and to the hot water tank to provide hot water from the taps in the house. Pump motors have to be switched on to direct the hot water to either, or both of, the radiator or hot water systems. The directing of water along these paths is determined by thermostats, one being in a room and activated by the air temperature and the other attached to the hot water tank for the hot water temperature. The entire process is to be set to operate only during certain hours of the day.

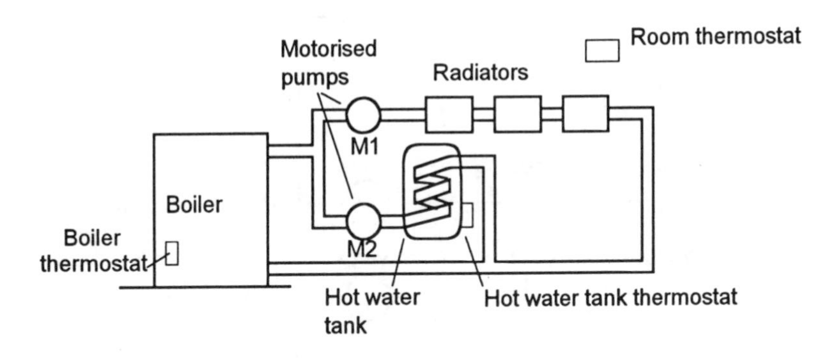

Figure 8.24 *Central heating system*

Figure 8.25 shows how a Mitsubishi PLC might be connected. The inputs are a clock-activated switch to X400. The input X401 is to switch on the radiator system and the input X402 to switch on the hot water system. The input X403 is the thermostat activated switch for the boiler, input X404 the air thermostat and input X405 the hot water tank thermostat. The outputs are Y430 to the boiler heater, Y431 to the pump M1 and Y432 to the pump M2.

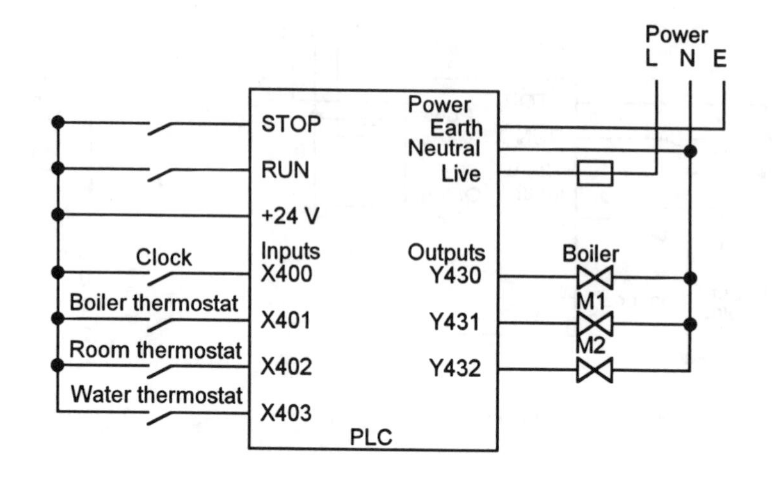

Figure 8.25 *PLC connections*

Figure 8.26 shows the ladder diagram. The boiler, output Y430, is switched on if X400 and X401 and either X402 or X403 are switched on. The valve M1, output Y431, is switched on if Y430 is on and X402 is on. The valve M2, output Y432,is switched on if Y430 is on and X403 is on.

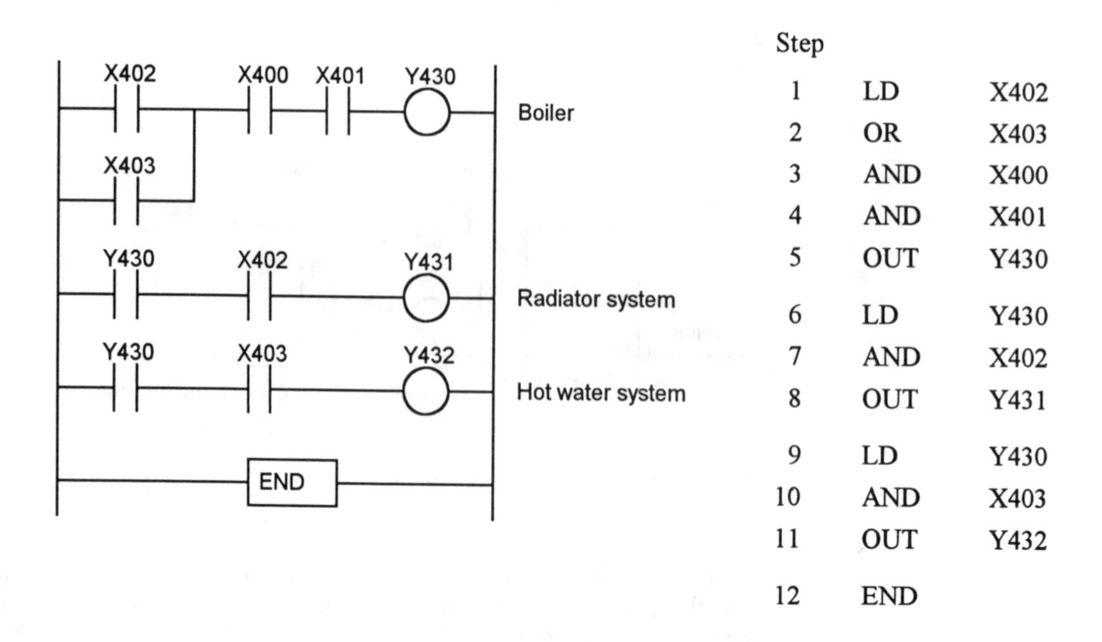

Step		
1	LD	X402
2	OR	X403
3	AND	X400
4	AND	X401
5	OUT	Y430
6	LD	Y430
7	AND	X402
8	OUT	Y431
9	LD	Y430
10	AND	X403
11	OUT	Y432
12	END	

Figure 8.26 *The ladder program*

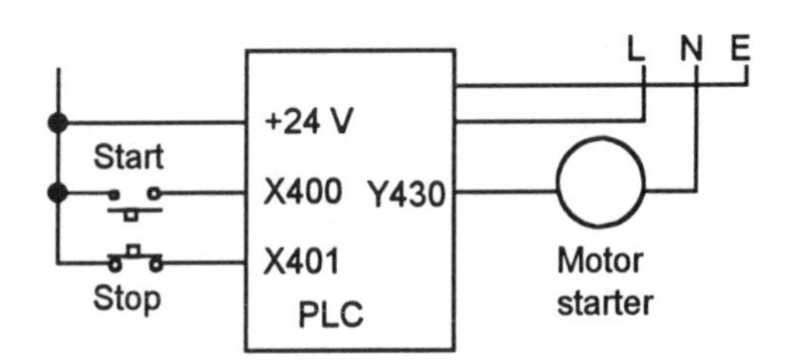

Figure 8.27 *Motor stop/start control system*

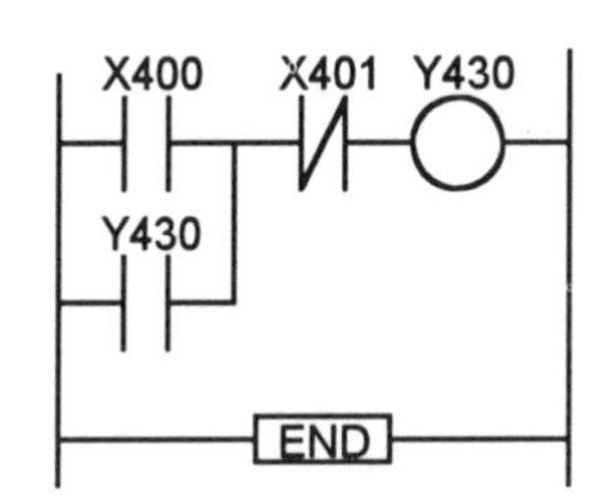

Figure 8.28 *Ladder diagram for motor*

Consider another example, a system which will start a motor when a push-button switch is pressed and will keep the motor running when the button is released, and then will stop the motor when another push-button switch is pressed, the motor remaining stopped even when the button is released. Figure 8.27 shows the basic elements of such system. What is required is a latching circuit (see Figure 6.44 and associated text). Figure 8.28 shows the required form of ladder diagram. X400 is the input for the start switch which is normally open. X401 is the input for the stop switch which is normally closed. When the start switch is closed, there is an output from Y430 and the motor starts. But associated with Y430 is a set of contacts which close when Y430 is activated. We thus now have an OR system between X400 and the Y430 contacts. Thus the output from Y430 continues. When the stop button is pressed, then the circuit is broken and so Y430 stops. Then the Y430 contacts open and so, even when contact X401 reverts to being closed, the motor remains off.

Consider another application, an automated drilling machine (Figure 8.29). The drill can be lowered or raised by means of a hydraulic cylinder activated by a solenoid-operated valve. The limit switch 1 is actuated when the drill is in the up position and not ready for drilling a hole. Before drilling can occur, the workpiece has to be clamped. This is achieved by means of a cylinder activated by a solenoid operate valve. The limit switch 3 is actuated when the workpiece is not clamped, limit switch 4 being actuated when it is clamped. When the workpiece is clamped and when the drill is lowered, then drilling occurs until limit switch 2 is actuated to indicate that the drill has drilled right through the workpiece. Then the drill is retracted and returned to the up position. Note that limit switches generally have two sets of contacts and so can be used to be normally open or normally closed.

We can break this sequence of operations down into a number of stages. Firstly we might have the drill motor and the hydraulic pump switched on. This could use a latching system similar to Figures 8.27 and 8.28. The next stage is probably to clamp the work piece. This means switching on solenoid 3 and keeping it on until the drill is retracted after drilling is completed. When solenoid 3 is switched on, limit switch 3 changes from closed to open and limit switch 4 changes from open to close. The next stage is to lower the drill. This means switching on solenoid 1. When the drill lowers, limit switch 1 changes from open to closed. The drill continues being lowered until limit switch 2 changes from closed to open. Then solenoid 1 switches off and solenoid 2 is switched on. Solenoid 2 remains switched on until limit switch 1 changes from closed to open. Figure 8.30 shows the input and output connects to a PLC and Figure 8.31 the resulting ladder diagram.

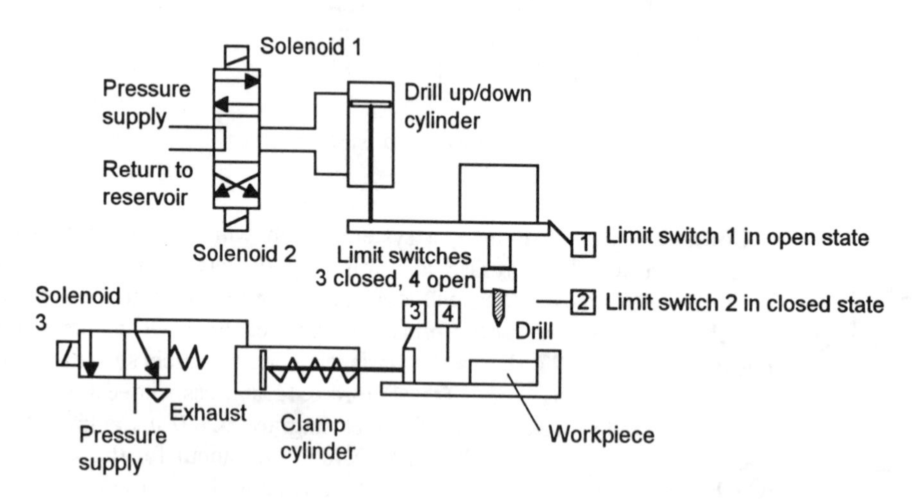

Figure 8.29 *The drilling machine*

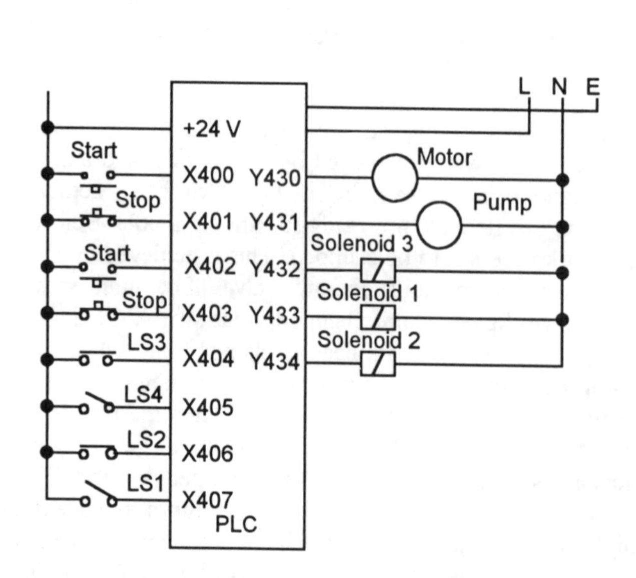

Figure 8.30 *PLC connections*

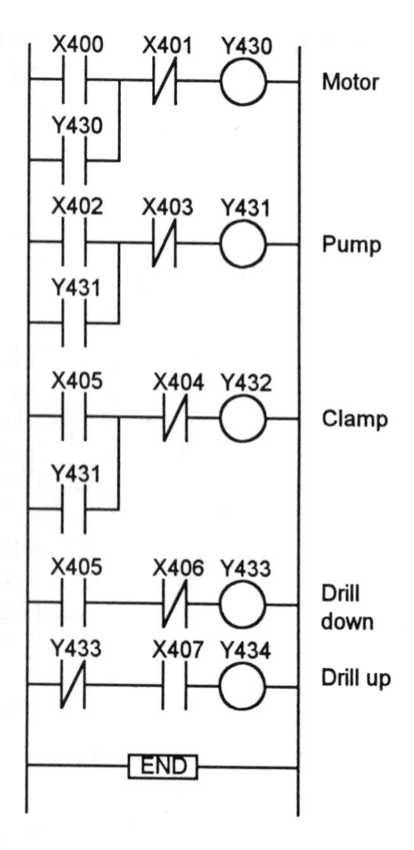

Figure 8.31 *Ladder diagram*

8.3.1 Using a timer

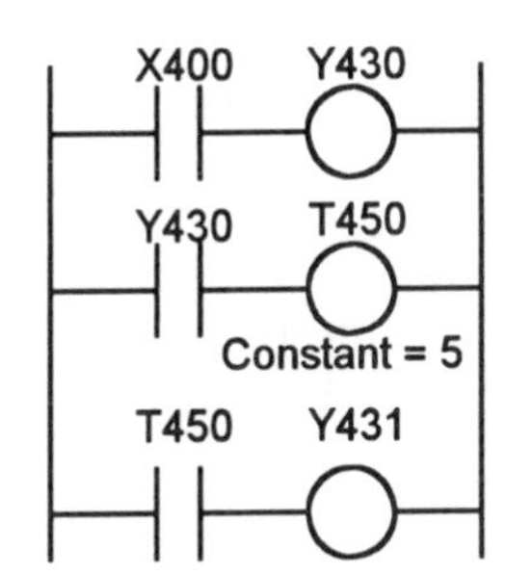

Figure 8.32 *Timer program*

As an example of the use of the timer, consider an application which requires two outputs in sequence so that the second output is switched on 5 s after the first output comes on. The ladder diagram will be like that shown in Figure 8.32. When X400 switches on then the output Y430 is actuated. This closes the contacts Y430 and starts the timer T450. The value of the timer constant is set to a value to give the time delay of 5 s. Then contacts T450 close and output Y431 is actuated.

There is often the need for an *on/off cycle timer* to make an output go on for some time, then off for some time, then on, and so on. Figure 8.33 shows a ladder diagram that can be used. When the X400 contacts close then timer T450 is started, the contacts for timer T451 being in the normally closed state when that timer is not running. After the required time, then the T450 contacts close and start the timer T451, also starting output Y430. After the time set for that timer, the contacts T451 open. This then switches off the timer T450 and so the output Y430. As a result of T450 contacts opening, time T451 switches off. Its contacts, which are then open, close. This then starts the cycle all over again.

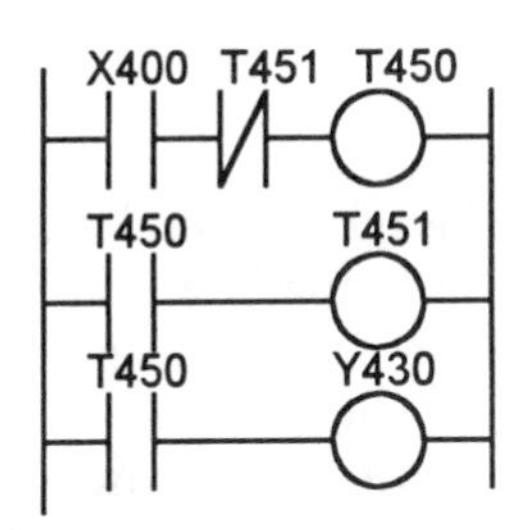

Figure 8.33 *On/off timer*

Another application which is often required is to switch an output off after some time delay, a so termed *on-delay timer*. Figure 8.34 shows the ladder diagram for such an application. When the contacts X400 are closed then the output Y430 is switched on and the timer Y450 started. After the time set for the timer, its contacts open, being normally closed. This then switches off the output Y430.

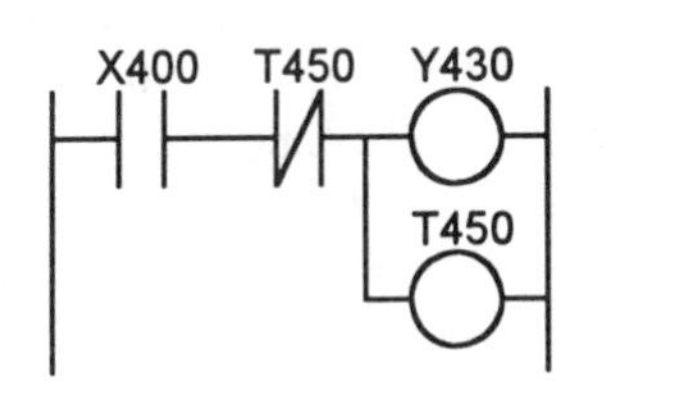

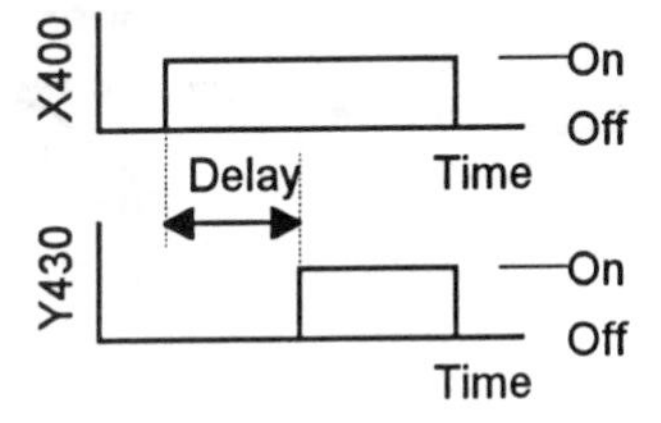

Figure 8.34 *On-delay timing*

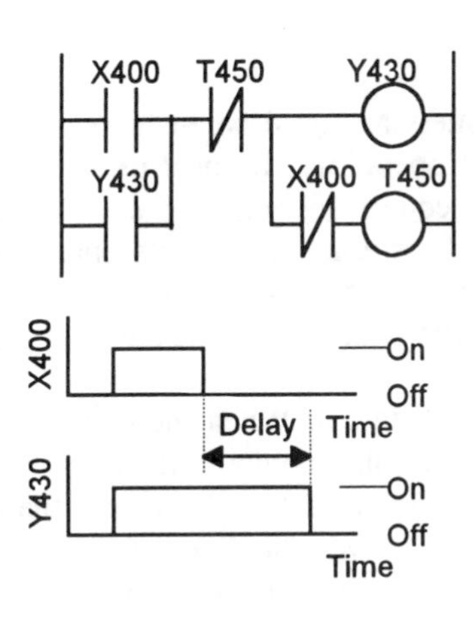

Figure 8.35 *Off-delay timing*

Figure 8.35 shows the ladder diagram for an *off-delay timer*. This can be used to introduce a delay between the stopping of an item and the stopping of another. When the contacts for the initial start switch X400 close the output Y430 starts. The timer cannot start because the normally closed X400 contacts in its branch open when the initial start switch X400 opens. When the initial start contacts X400 open, Y430 continues because the X400 is latched. But the opening of the initial start contacts closes the normally closed X400 contacts and so the timer starts. When the timer period is up and it opens the T450 contacts, the output Y430 is switched off.

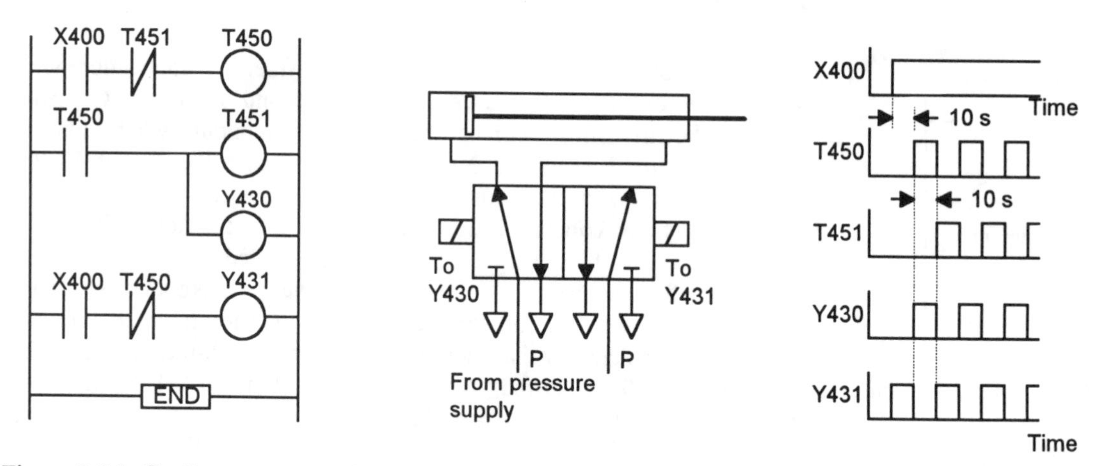

Figure 8.36 *Cyclic movement of a piston*

As an illustration of the use that can be made of a timer, consider the circuit that can be used for the cyclic movement of a piston in a cylinder (Figure 8.36). This piston might be used to periodically push work pieces into position in a machine tool, another similar but out of phase arrangement being used to remove completed workpieces. Suppose the timers are each set for times of 10 s. The two timers set and reset each other every 10 s. When the start contacts X400 are closed, timer T450 starts. The T450 contacts for the solenoid connected to Y431 are in the normal state of closed and so that solenoid is energised and the piston is moved to the left. When the 10 s is up, the T450 contacts close. This energises the solenoid connected to Y430 and causes the piston to move to the right. At the same time, the timer T431 is started. After 10 s the T451 contacts close and timer T450 starts. So the entire sequence repeats itself.

8.3.2 Using a counter

PLCs have some inputs for counters. Counters are used when there is a need to count a specified number of contact operations. In most cases the counter operates as a *down counter* in that it counts down from the present value to zero. When the zero is reached, the counter's contacts change state. An *up-counter* counts from zero upwards. Counters are used for such applications as counting the number of revolutions of a shaft, the number of items passing along a conveyor belt, etc.

Figure 8.37 shows a basic counting ladder diagram. When the contacts X400 momentarily close, the counter resets and will then count the number of pulses from the X401 contacts. Each time the contacts X401 close, the counter counts 1. When the count for which the counter has been set, in the example given it is 10, is reached the contacts C460 close and so output Y430 is activated. If at any time the contacts X400 are closed, the counter will resets and again start counting from 10.

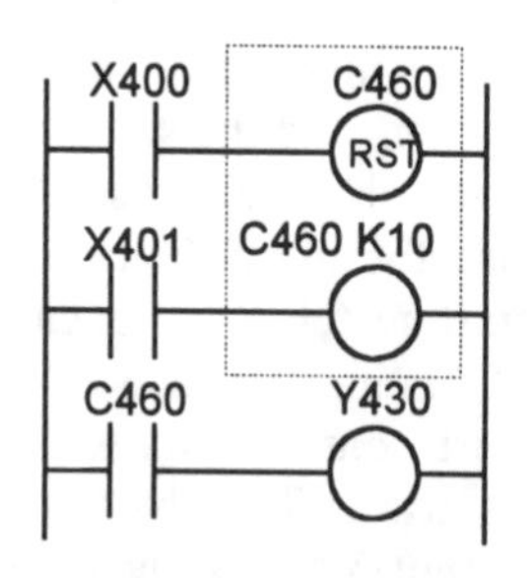

Figure 8.37 *Counter ladder diagram*

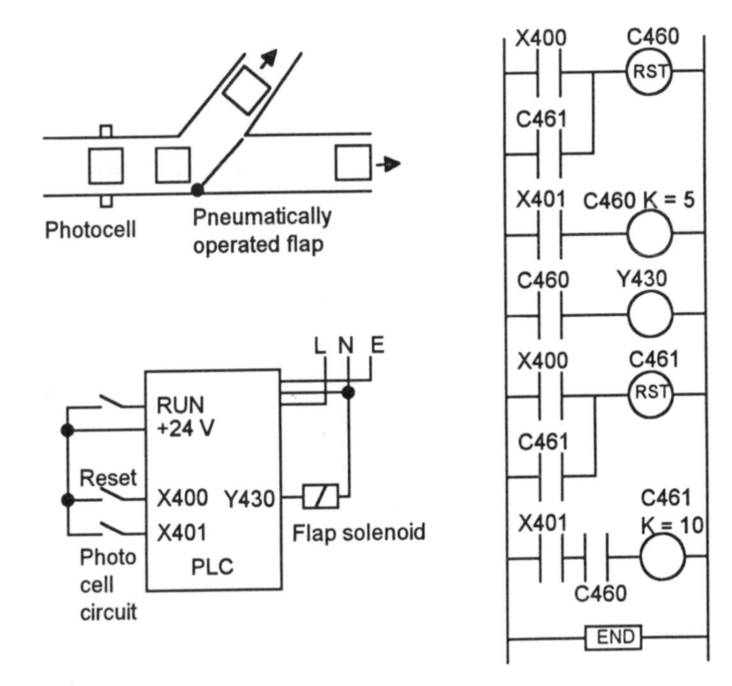

Figure 8.38 *Batch counting*

To illustrate the use of a counter, consider a production situation where the number of items being diverted from a conveyor and the number continuing along it are to be controlled (Figure 8.38). Thus we might have a pneumatically operated flap which is activated to divert items and which is activated to divert 5 items and then allow the next 10 items to continue. The signal for the counter is obtained by a photoelectric cell giving an output each time a beam of light is interrupted by an item passing. When X400 contacts are momentarily closed, counters C460 and C461 are both reset, rungs 1 and 4 of the ladder. The photocell counts until a count of 5 is reached. Then C461 closes its contacts. This then activates Y430 and the solenoid opens the flap to allow items to pass on. It also starts C461 counting, rung 5 of the ladder. When 10 items have been counted, the counters are both reset and the operation starts again.

8.3.3 Using an internal relay

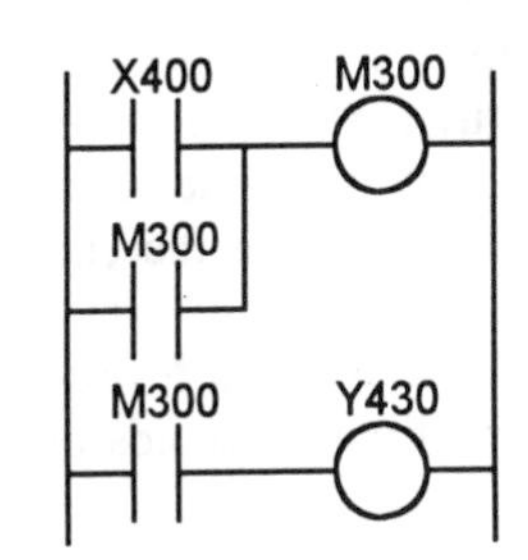

Figure 8.39 *Power failure system*

The term *marker* or *auxiliary relay* or *internal relay* is used for an element which is similar to an output but is internal within the PLC and behaves like a relay. It can be used to store intermediate results.

Figure 8.39 illustrates this behaviour by considering a simple application of using a battery-backed internal relay for a system to cope with a power failure which would result in the input to a PLC switching off and so, as a consequence, change an output to some device. When the contacts X400 close the coil of the internal relay M300 is energised. This closes the M300 contacts and, since they are in an OR situation with X400, even if X400 opens as a result of power failure the M300 contact remains closed. Thus the output Y430 remains on.

8.3.4 Using the shift register

The term *register* is used for an electronic device in which data can be stored. The shift register is a number of internal relays grouped together to form a register. Each internal relay is either effectively shut or open, these states being designated by a 0 and a 1. The term *bit* is used for such a binary digit. Thus if we have four internal relays in the register we can store four 0/1 states, i.e. it is a 4-bit register. With the shift register it is possible to shift stored bits. Thus we might have, for the 4-bit register,

1	0	1	0

If we then have an input of a 1 shift pulse then we obtain

Input → | 1 | 1 | 0 | 1 | Overflow of 0

All the bits have shifted along one place and the last bit has overflowed.

Shift registers require at least three inputs, one to load data into the first element of the register, one as the shift command and one for resetting, i.e. clearing the entire register. To illustrate the application of a four-bit shift register, consider the ladder diagram given in Figure 8.40. M140 has been designated as the first internal relay, M141 as the second, M142 as the third and M143 as the fourth. When X400 closes, the contents of the shift register are set to 0 (RST = reset). When X401 is momentarily closed there is a 1 output to the M140 internal relay. The register then reads 1 0 0 0. The M140 contacts close and so there is an output from Y430. When X402 momentarily closes, the pulse in the first relay moves on to the second relay. The register then reads 0 1 0 0. This means that M141 closes and so there is an output to Y431, but now there is no output from Y430 since M140 is closed. When X402 again is momentarily closed, the pulse in the second relay is moved to the third relay. The register then reads 0 0 1 0. This means that M142 closes and so there is an output to Y432, the other outputs all being off. When X402 again is momentarily closed, the pulse in the third relay moves to the fourth relay. The register then reads 0 0 0 1. This means that M143 closes and so there is an output to Y433, the other outputs all being off. When X402 again is momentarily closed, the pulse in the fourth relay overflows and so all the registers are now 0. All the outputs are then off. When X400 is now momentarily closed we have the entire cycle repeating itself.

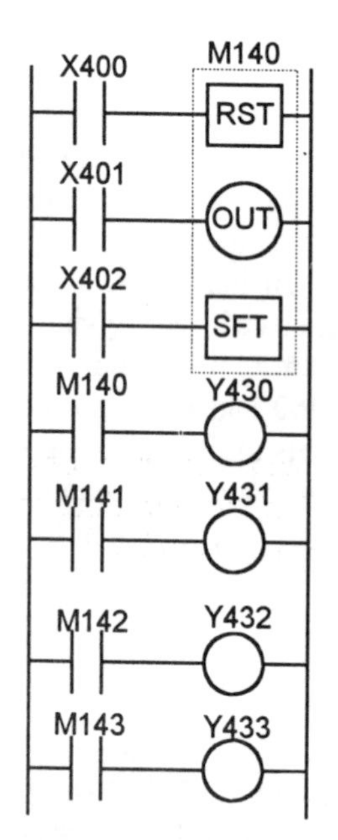

Figure 8.40 *Shift register*

Shift registers are thus used where a sequence of operations is required. This might, for example, be the sequential operation of the solenoids of a number of pneumatic valves.

8.3.5 Using the master control relay

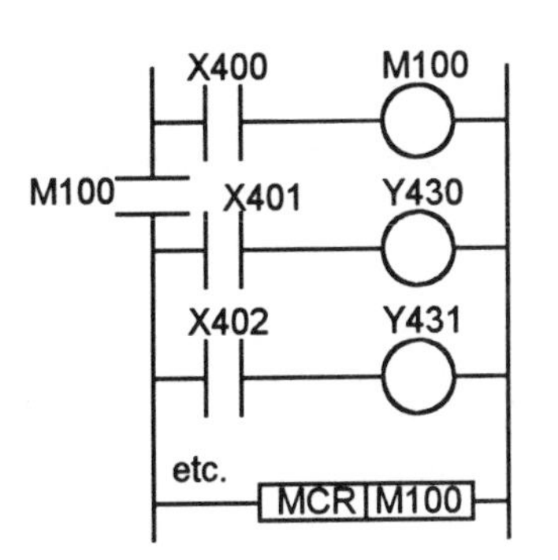

Figure 8.41 *Using the master control relay*

When large numbers of outputs are being controlled it is sometimes desirable for whole sections of the ladder diagram to be turned off or on when certain criteria are realised. This could be done by including the same internal relay in each of the rungs so that operation of it affects all of them. An alternative is to use a master control relay to control all those rungs. Figure 8.41 illustrates this. M100 is given the program instruction to be a master control relay. With Mitsubishi this means entering the code MC M100. This means that effectively the contacts for M100 are inserted in the power line to succeeding rungs in the way shown. Until X400 closes and gives an output from the internal relay M100, there is no power to the succeeding rungs and so Y430, Y41, etc., cannot be switched on, regardless of the conditions of the inputs X401, X402, etc. The end of a master control section is signified by either the occurrence of another master control line or by programming a reset for the master control relay. With Mitsubishi this means entering the code MCR M100. The instruction code for Figure 8.41 is:

```
1    LD     X400
2    OUT    M100
3    MC     M100
4    LD     X401
5    OUT    Y430
6    LD     X402
7    OUT    Y431
etc. up to the reset at perhaps step 12.
12   MCR    M100
```

Such a master control relay might be added to the ladder diagram for the drilling machine described by Figure 8.29 so that none of the operations will function if the machine guard is not in position and closing its switch.

8.3.6 Using conditional jumps

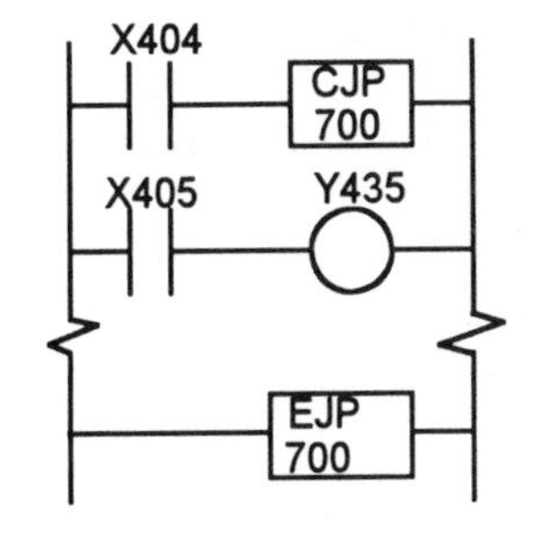

Figure 8.42 *Conditional jump*

Situations can occur where one set of actions needs to occur if one condition occurs and another set if is does not occur. Thus if the condition is met there needs to be a jump to a different part of the program. Figure 8.42 shows how this can be achieved using the conditional jump. If X404 closes then the conditional jump 700 (CJP = conditional jump) is activated and the program skips all the intermediate steps and goes to the line with EJP700 (EJP = end conditional jump) and then continues with the program that follows it. If X404 is not closed then the program continues with the next line, i.e. the one starting with X405.

Problems

Questions 1 to 12 have four answer options: A, B, C and D. Choose the correct answer from the answer options.

1 Decide whether each of these statements is True (T) of False (F).

Velocity feedback is used with a servo system for controlling the position of a load because:
(i) It eliminates the steady state error.
(ii) It gives damping.

A (i) T (ii) T
B (i) T (ii) F
C (i) F (ii) T
D (i) F (ii) F

Questions 2 to 6 refer to Figure 8.43 which show a system used to control the position of a load.

2 The feedback from the optical encoder is used to give:

A An analogue signal related to the position of the load.
B An analogue signal related to the speed of the load.
C A digital signal related to the position of the load.
D A digital signal related to the speed of the load.

3 The feedback from the tachogenerator is used to give:

A An analogue signal related to the position of the load.
B An analogue signal related to the speed of the load.
C A digital signal related to the position of the load.
D A digital signal related to the speed of the load.

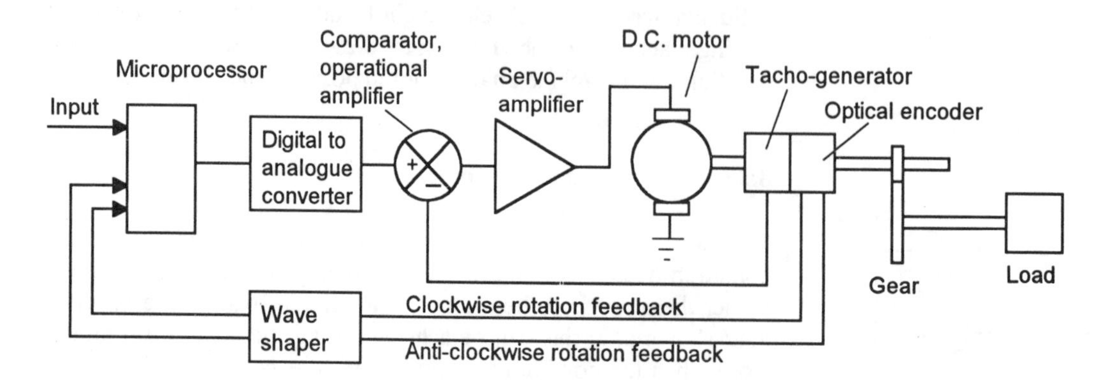

Figure 8.43 *System used to control the position of a load*

4 Decide whether each of these statements is True (T) of False (F).

The digital to analogue converter is included because:
(i) The microprocessor gives a digital output.
(ii) Operational amplifiers require analogue inputs.

A (i) T (ii) T
B (i) T (ii) F
C (i) F (ii) T
D (i) F (ii) F

5 Decide whether each of these statements is True (T) of False (F).

The feedback from the tachogenerator is to provide:
(i) Damping and hence stabilisation.
(ii) Feedback of the position of the load.

A (i) T (ii) T
B (i) T (ii) F
C (i) F (ii) T
D (i) F (ii) F

6 Decide whether each of these statements is True (T) of False (F).

A d.c. motor is shown with the error current fed to the armature coils. Then:
(i) The angle through which the motor shaft rotates is proportional to the error.
(ii) The speed with which the motor shaft rotates is proportional to the error.

A (i) T (ii) T
B (i) T (ii) F
C (i) F (ii) T
D (i) F (ii) F

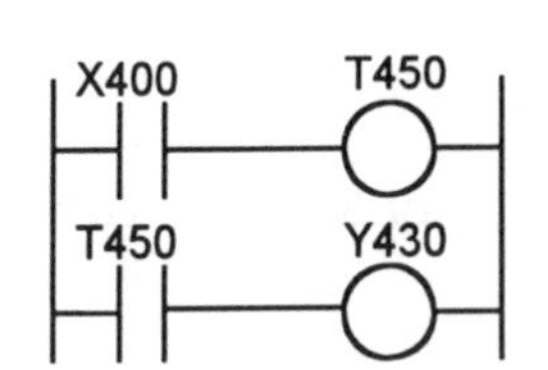

Figure 8.44 *Ladder diagram*

7 Figure 8.44 shows a ladder diagram involving a timer set to give a time interval of 5 s. When X400 is activated and kept closed, the output Y430 is:

A Immediately activated and remains switched on.
B Activated after 5 s and remains switched on.
C Immediately activated and switched off after 5 s.
D Activated after 5 s and then switched off after a further 5 s.

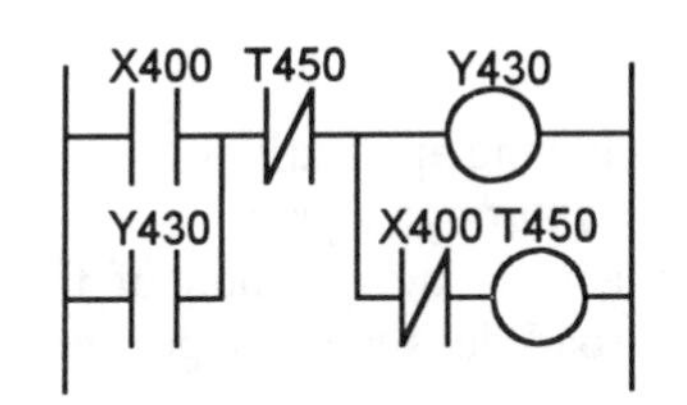

Figure 8.45 *Ladder diagram*

8 Figure 8.45 shows a ladder diagram involving a timer set to give an interval of 5 s. When X400 is activated, the output Y430 is:

A Immediately activated and remains on for as long as X400 is on.
B Activated after 5 s and remains on, even when X400 switched off.
C Immediately activated and switched off 5 s after X400 switched off.
D Activated after 5 s and switched off when X400 switched off.

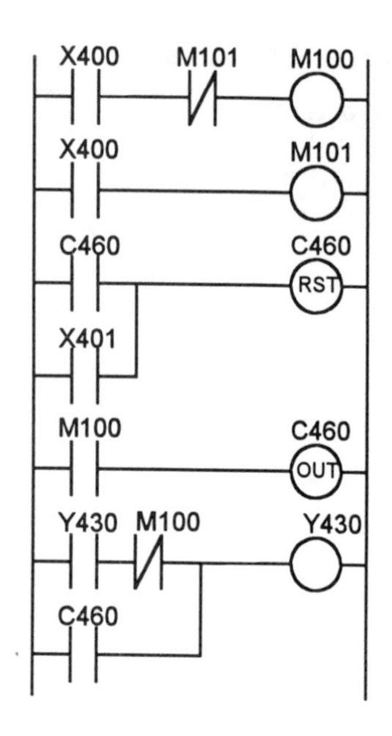

Figure 8.46 *Ladder diagram*

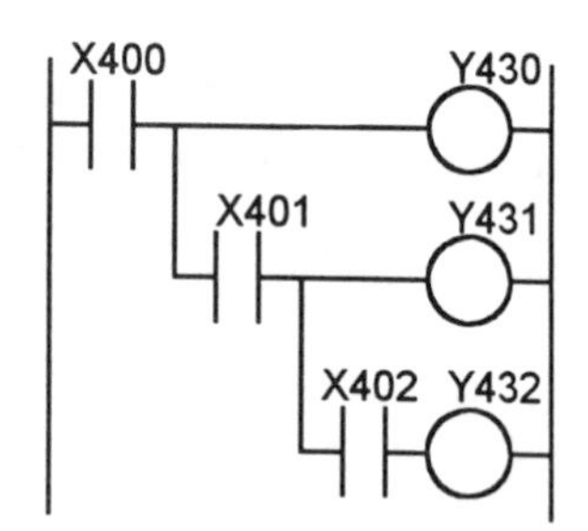

Figure 8.47 *Ladder diagram*

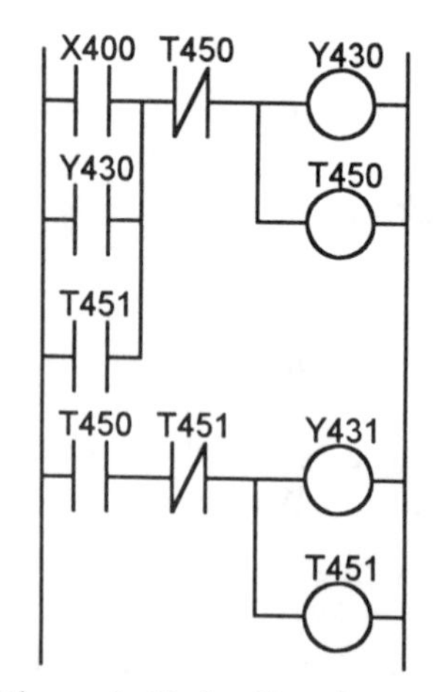

Figure 8.48 *Ladder diagram*

Questions 9 and 10 refer to Figure 8.46 which shows a ladder circuit involving a counter C460 set to count down from 10 to 0 and two internal relays M100 and M101.

9 When X400 is closed and kept closed, the contacts of the internal relay M100:

A Remain constantly open on as long as X400 is closed.
B Remain constantly closed as long as X400 is closed.
C Open for a short duration.
D Close for a short duration pulse.

10 When X405 is closed and opened for 10 times, the output Y430 is:

A Activated on the tenth pulse and remains on until the next pulse.
B Activated on the tenth pulse and remains on for the next ten pulses.
B Switched off on the tenth pulse.
C Has been activated ten times.

11 Decide whether each of these statements is True (T) of False (F).

Figure 8.47 shows a ladder diagram which has three inputs X400, X401 and X402, together with three output Y430, Y431 and Y432.
(i) To switch on output Y400, only X400 needs to be closed.
(ii) To switch on output Y403, X400, X401 and X402 must all be closed.

A (i) T (ii) T
B (i) T (ii) F
C (i) F (ii) T
D (i) F (ii) F

12 Decide whether each of these statements is True (T) of False (F).

Figure 8.48 shows a ladder diagram with timers T450 and T451, each set for 5 s. Following the closure of X400:
(i) Output Y430 will be on for 5 s, followed by off for 5 s, followed by on for 5 s, and so on.
(ii) Output Y431 will be off for 5 s, followed by on for 5 s, followed by off for 5 s, and so on.

A (i) T (ii) T
B (i) T (ii) F
C (i) F (ii) T
D (i) F (ii) F

13 Devise systems which can be used for the following applications:
(a) To control the rate at which a liquid passes through a pipe.
(b) A car headlamp levelling system that adjusts the position of the lights so that the angle of the beams to the road does not change when the load in the vehicle changes.

(c) A machine which will control the components moving along a conveyor belt and reject those that are faulty, e.g. bottles without caps.
(d) A small robot arm which is to be controlled to have an upward or downward movement.
(e) To control the temperature of a car engine by adjusting the rate of flow of coolant.
(f) To control the speed of revolution of the shaft of a d.c. motor.
(g) A machine tool to machine a workpiece to a scaled up version of a template.
(h) A device which will position a workpiece on the table of a machine tool so that it is at the right angle and in the right position.

Answers

The following are answers to the numerical and multiple-choice problems and brief clues as to the answers to other problems.

Chapter 1

1 A 2 B 3 B 4 C 5 B 6 D
7 A

8 Sensor, signal processor, data presentation, see section 1.2.
9 See sections 1.4.2 and 1.4.3. Reliability: the probability that a system will operate to an agreed level of performance for a specified period, subject to specified environmental conditions. Repeatability: the ability of a system to give the same value for repeated measurements of the same variable.
10 See section 1.4.1.
11 See Section 1.4.
12 Only give required accuracy 6 times in 10.
13 In 1 year, 1 in 100 will be found to fail.

Chapter 2

1 C 2 B 3 A 4 C 5 B 6 A
7 B 8 B 9 D 10 B 11 C 12 C
13 A 14 C 15 A 16 C 17 B 18 C
19 A 20 A

21 For example: (a) turbine meter, (b) bourdon gauge, (c) LVDT, (d) thermocouple.
22 For example: (a) Wheatstone bridge, (b) ADC, (c) multiplexer.

Chapter 3

1 A 2 A 3 C 4 C 5 D

6 For example: painted stripes on floor and optical sensor; optical encoder to monitor rotation of wheel.
7 For example: (a) weight-operated electrical switch, (b) thermistor, potential divider circuit, voltmeter, (c) load cell using strain gauges, Wheatstone bridge, amplifier.

Chapter 4

1 A 2 A 3 C 4 A

5 See Figure A.1.
6 See section 4.2.

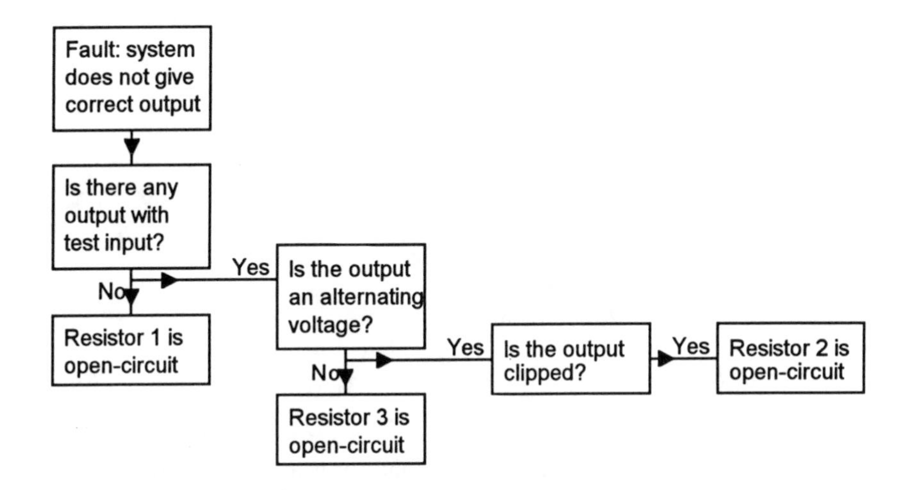

Figure A.1 *Problem 5 Chapter 4*

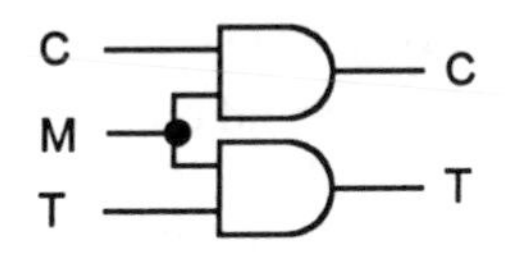

Figure A.2 *Problem 16 Chapter 5*

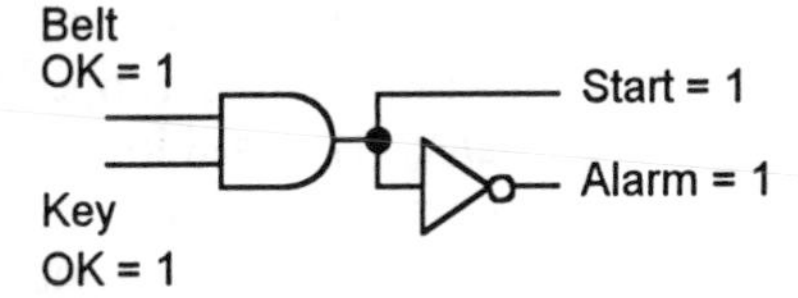

Figure A.3 *Problem 17 Chapter 5*

Chapter 5

1 D	2 A	3 A	4 D	5 B	6 C
7 A	8 D	9 B	10 A	11 A	12 B
13 B					

14 (a) Closed-loop, (b) sequential, (c) NOT gate, (d) Open-loop or perhaps closed-loop, (e) closed-loop, (f) open-loop, (g) closed-loop.
15 (a) Closed-loop with on-off output from controller, (b) direct digital control with digital input and digitial output.
16 See Figure A.2.
17 See Figure A.3

Chapter 6

1 C	2 A	3 B	4 D	5 A	6 C
7 B	8 D	9 A	10 C	11 C	12 A
13 C	14 B	15 C	16 A	17 A	18 B
19 B	20 D				

21 See sections 6.3 to 6.6.
22 See section 6.7
23 See (a) Figure 6.6, (b) section 6.8.1, (c) section 6.8.2.

Chapter 7

1 B	2 C	3 A	4 D	5 A	6 A
7 B	8 C	9 D	10 A	11 A	12 C
13 D	14 B	15 A	16 B	17 A	18 C
19 C	20 A				

21 0.004 m^2
22 5.85 m^3/s
23 504 pulses/s
24 1260 mm
25 800 mm

Chapter 8

1 C	2 C	3 B	4 A	5 B	6 C
7 B	8 C	9 D	10 A	11 A	12 A

13 The systems might, for example, include the following items: (a) closed loop, orifice plate, differential pressure sensor, diaphragm operated flow control valve; (b) closed loop, sensors for the suspension movements, control unit, actuators to tilt lamps; (c) PLC, photoelectric sensor to give a pulse input, output to solenoid actuated lever; (d) PLC, input of limit switches, output to solenoid operated valve; (e) closed loop, thermistor sensor, operational amplifier comparator, solenoid operated valve; (f) closed loop, tachogenerator sensor, operational amplifier comparator, servo amplifier (as part of Figure 8.16); (g) template to move input potentiometer slider and system then as in Figure 8.16 or one of the later Figures; (h) PLC, solenoid operated valve controlling cylinder and limit switches for the *x*-direction and for the *y*-direction, see Figure 8.29 for an example of such a positioning device.

Index